Albert Michelson's Harmonic Analyzer

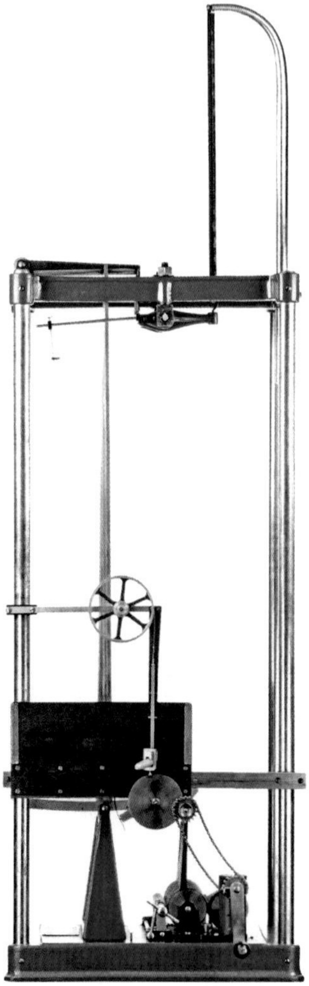

0°

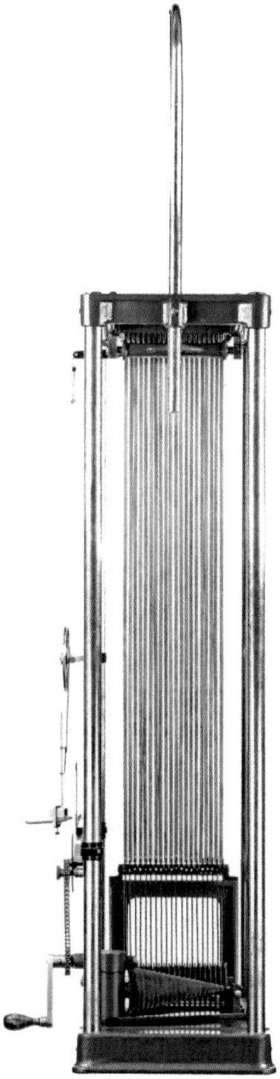

90°

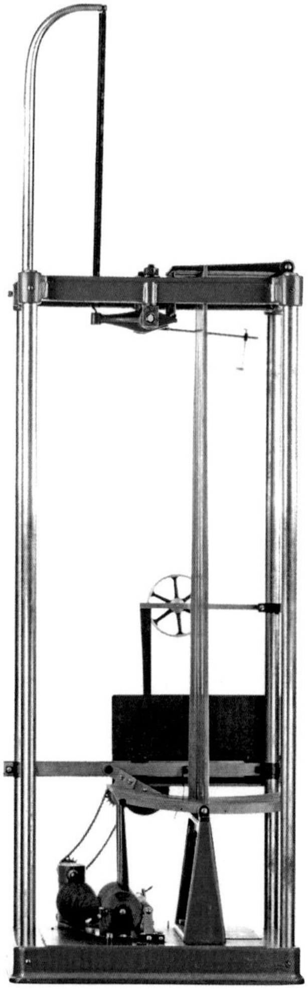

180°

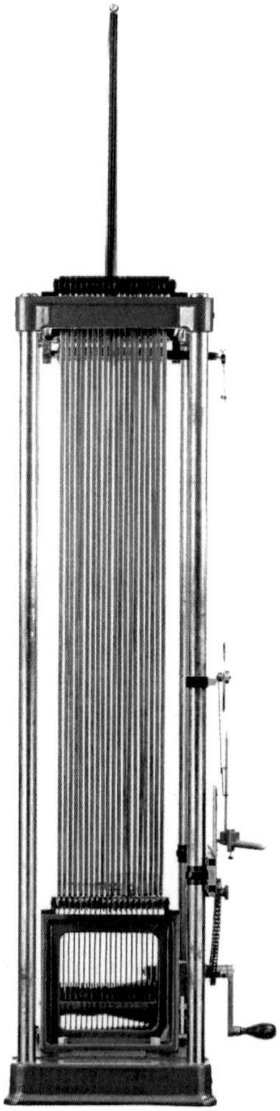

270°

Albert Michelson's

Harmonic Analyzer

A Visual Tour of a Nineteenth Century Machine that Performs Fourier Analysis

Bill Hammack, Steve Kranz & Bruce Carpenter

Articulate Noise Books

Albert Michelson's Harmonic Analyzer:
A Visual Tour of a Nineteenth Century Machine that Performs Fourier Analysis

Bill Hammack, Steve Kranz, Bruce Carpenter
ISBN-13: 978-0-9839661-6-6 (hardcover)
ISBN-13: 978-0-9839661-7-3 (paperback)
1. Mathematical instruments—Calculators 2. Michelson, Albert A. (Albert Abraham), 1852–1931
3. Mathematical analysis—Fourier analysis—Fourier Series

Photography and design by Steve Kranz

Epigram

To Michelson:

What manner of man was so wise,
As to make a machine Synthesize?
  —With springs and levers it combines
    Weighted sines or cosines—
And most wondrous of all: Analyze!

Contents

Preface	ix
Introduction	1
Visual Table of Contents	4
Fourier Synthesis	6
Fourier Analysis	8
The Harmonic Analyzer	11
Crank	12
Cone Gear Set	16
Cylinder Gear Set	22
Rocker Arms	26
Amplitude Bars	30
Measuring Stick	34
Springs and Levers	38
Summing Lever	42
Counter Spring	44
Magnifying Lever	46
Magnifying Wheel	50
Platen	54
Translational Gearing	56
Pen Mechanism	64
Pinion Gear	66
Provenance	70
Output from the Machine	75
Michelson's 1898 Paper	90
Math Overview	98
Eight Views of the Machine	101
Notes on the Design	118

Preface

IN OCTOBER 2012 THE THREE of us set out to create a video series to illuminate the hidden importance of Fourier methods in our modern technological world. We had intended to stay firmly rooted in the twenty-first century; instead we discovered a machine that took us over one hundred years into the past.

We learned, while researching Fourier methods, of nineteenth century machines that performed Fourier synthesis and analysis. We thought such a machine would be an ideal subject for a video series to present Fourier methods in a highly visual way. This line of thought awoke in two of us—Bill and Bruce—dim memories of such a machine located somewhere in Altgeld Hall, home of the University of Illinois's Department of Mathematics.

We rushed to that building, a three-minute walk from where we were planning our video series, and found, sitting in a glass case in the second floor hallway, a wonderful contraption of gears, spring and levers—a Fourier analyzer. The Department of Mathematics graciously granted our request to free the analyzer from its case so that we could film it. We moved it to our machine shop, where Mike Harland and Tom Wilson designed and built a replacement for the missing mechanism that holds the pen. We thank the members of our Advance Reader Program for their very useful comments and corrections.

We brought the machine into our studio, and as we investigated its operation, its charms overwhelmed us. It became the star of the video series, and the subject of this book. While the book stands alone, we encourage readers to watch the videos exploring its operation at *www.engineerguy.com/fourier.*

Bill Hammack, Steve Kranz & Bruce Carpenter

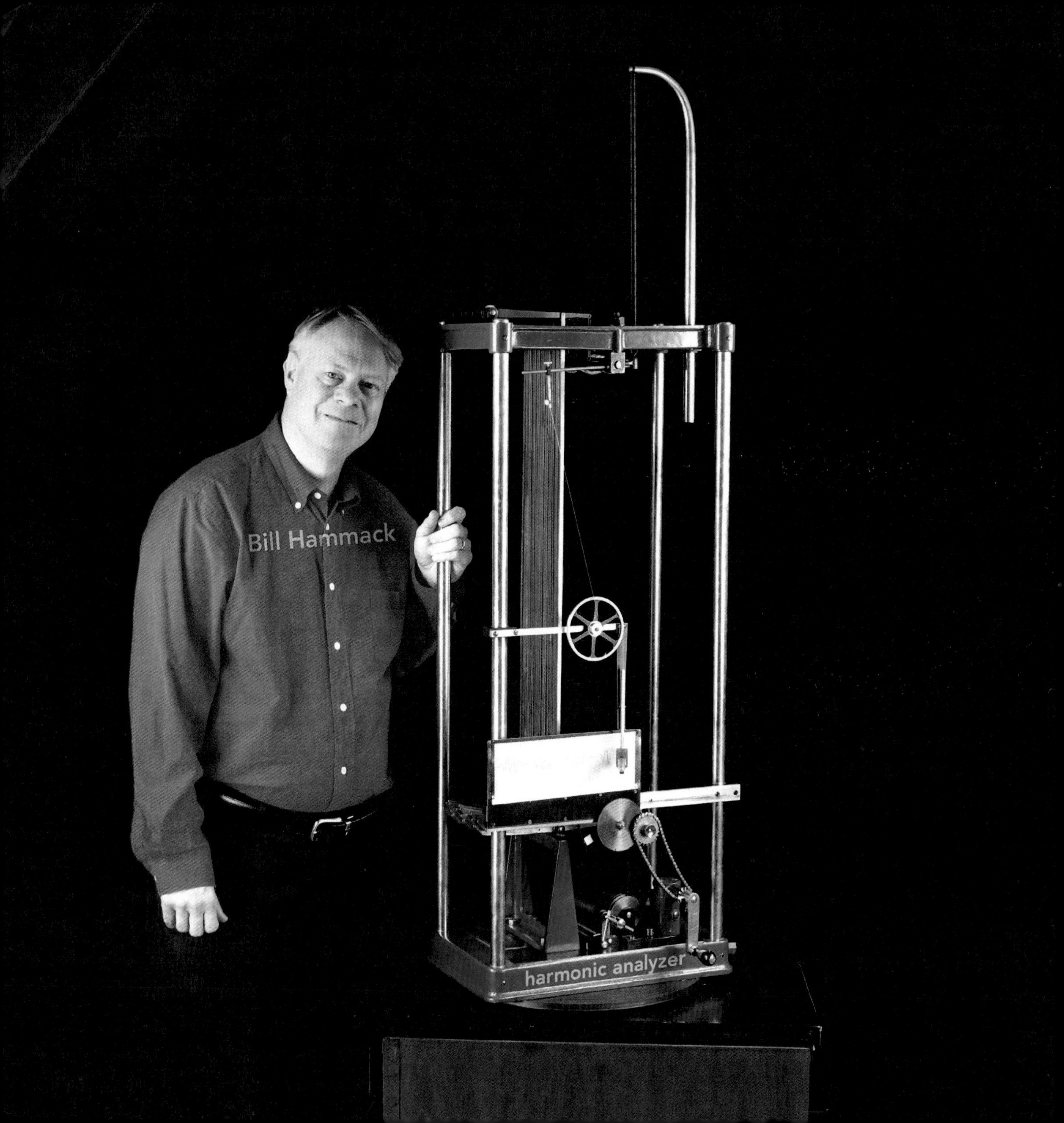
Bill Hammack
harmonic analyzer

Introduction

THIS BOOK CELEBRATES a harmonic analyzer designed in the late nineteenth century by the physicist Albert Michelson. A harmonic analyzer can carry out two related tasks: it can add together weighted sines or cosines to produce a function, and it can perform the inverse operation of decomposing a given function into its constituent sinusoids.

The addition of sinusoids is called *Fourier synthesis*. While adding only sinusoids seems limiting, the machine can create beautiful patterns that look nothing like sinusoids: it produces beats, peaks, flat sections, or other complicated patterns.

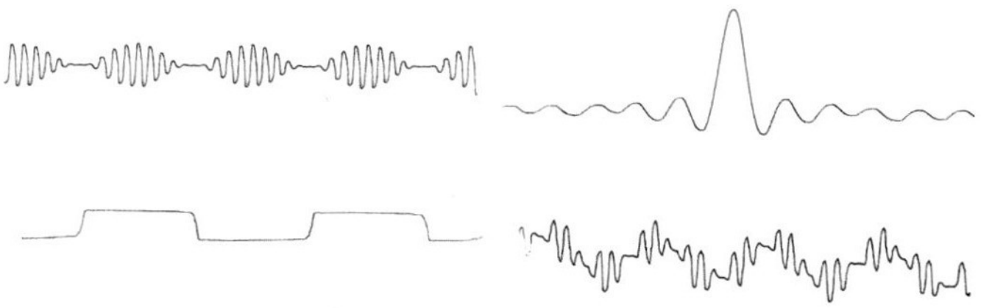

These patterns were produced by the harmonic analyzer described in this book. The pattern in the *upper left* is beats, *upper right* a sinc, and *lower left* a square wave. The pattern in the *lower right* was made by setting the machine's amplitudes bars randomly.

Astonishingly, the machine can also reveal the recipe for making these rich patterns. Given any even or odd periodic function, the analyzer can calculate the proper weighting to use when approximating that function by a series of cosines or sines. This mathematical operation is called *Fourier analysis*. A generalized form of Fourier analysis is central in solving many scientific and engineering problems. A few examples of their diverse applications include: removing noise from images sent by NASA space probes, compressing sound recordings to make MP3s, and determining the arrangement of atoms in a crystal.

Today, mention of Michelson brings to mind the Nobel Prize winning Michelson-Morley experiment, that famous measurement of the speed of light that ruled out a stationary light-bearing "luminiferous aether." Yet he studied many different physical phenomena, among them the light emitted by flames. He noted that a flame made by burning even a pure element was composed of light of different frequencies. Michelson wanted to know the exact values of these frequencies. He measured the emission from these elements using an interferometer, the same type of device he used in the Michelson-Morley experiment. In an interferometer a beam of light is split into two paths and then recombined. By varying the length of one of the beams Michelson could cause the recombined beams to interfere constructively or destructively. The amount of interference depended on the frequencies of light in the beam. To extract the frequencies he used Fourier analysis. At first Michelson performed by hand the Fourier analysis needed to determine those frequencies, but soon found it laborious. "Every one who has had occasion," he once wrote,

> "to calculate or to construct graphically the resultant of a large number of simple harmonic motions [sinusoids] has felt the need of some simple and fairly accurate machine which would save the considerable time and labor involved in such computations."

This need lead him to the invention and construction of the harmonic analyzer described in this book.

He began by studying the scientific literature on harmonic analyzers. He found only one "practical instrument": an analyzer developed by Lord Kelvin to calculate tide tables. To create the sinusoidal motions needed to simulate tides Kelvin strung ropes around pulleys. These ropes were the machine's great flaw, as Michelson, a superb experimentalist, immediately saw:

> "The range of the machine is however limited to a small number of elements on account of the stretch of the cord and its imperfect flexibility, so that with a considerable increase in the number of elements the accumulated errors due to these causes would soon neutralize the advantages of the increased number of terms in the series."

To eliminate the problems caused by the stretching ropes Michelson considered several solutions. "Among the methods which appeared most promising," he wrote, "were addition of fluid pressures, elastic and other forces, and electrical currents. Of these the simplest in practice is doubtless the addition of the forces of spiral springs."

Using springs he first built a 20-element analyzer, one that calculates with 20 sinusoids with radian frequencies starting at 1, the fundamental, followed by the harmonics 2, 3, and so on up to 20. He found the "results obtained were so encouraging that it was decided to apply to the Bache Fund for assistance in building the present machine of eighty elements." His application succeeded: he got $400.00. With those funds he built a harmonic analyzer with 80 elements, which he described in detail in an article published in *The American Journal of Science*. (This paper is reproduced on pg. 90.)

In that paper Michelson mentions his plans to build an analyzer with 1000 elements. His grand vision never came to be, perhaps because of technical limitations in materials and machining, or perhaps because of other demands on Michelson's time. And while this machine was never built, with today's computational power we essentially have Michelson's harmonic analyzer built into many devices: it is in every mobile phone, every telecommunications system, and in every computer program that manipulates an image.

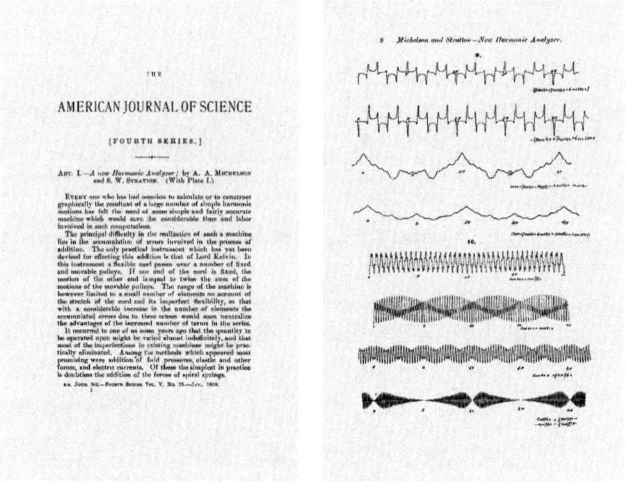

THE

AMERICAN JOURNAL OF SCIENCE

[FOURTH SERIES.]

ART. I.—*A new Harmonic Analyzer*; by A. A. MICHELSON and S. W. STRATTON. (With Plate I.)

EVERY one who has had occasion to calculate or to construct graphically the resultant of a large number of simple harmonic motions has felt the need of some simple and fairly accurate machine which would save the considerable time and labor involved in such computations.

The principal difficulty in the realization of such a machine lies in the accumulation of errors involved in the process of addition. The only practical instrument which has yet been devised for effecting this addition is that of Lord Kelvin. In this instrument a flexible cord passes over a number of fixed and movable pulleys. If one end of the cord is fixed, the motion of the other end is equal to twice the sum of the motions of the movable pulleys. The range of the machine is however limited to a small number of elements on account of the stretch of the cord and its imperfect flexibility, so that with a considerable increase in the number of elements the accumulated errors due to these causes would soon neutralize the advantages of the increased number of terms in the series.

It occurred to one of us some years ago that the quantity to be operated upon might be varied almost indefinitely, and that most of the imperfections in existing machines might be practically eliminated. Among the methods which appeared most promising were addition of fluid pressures, elastic and other forces, and electric currents. Of these the simplest in practice is doubtless the addition of the forces of spiral springs.

AM. JOUR. SCI.—FOURTH SERIES, VOL. V, NO. 25.—JAN., 1898.
1

In 1898 Michelson and his coauthor Stratton published a paper in *The American Journal of Science* that detailed an 80-element harmonic analyzer closely related to the 20-element analyzer featured in this book. A facsimile of the complete paper is included in this book (pg. 90).

Albert Michelson (1852-1931)

Dimensions

weight: 69 kg

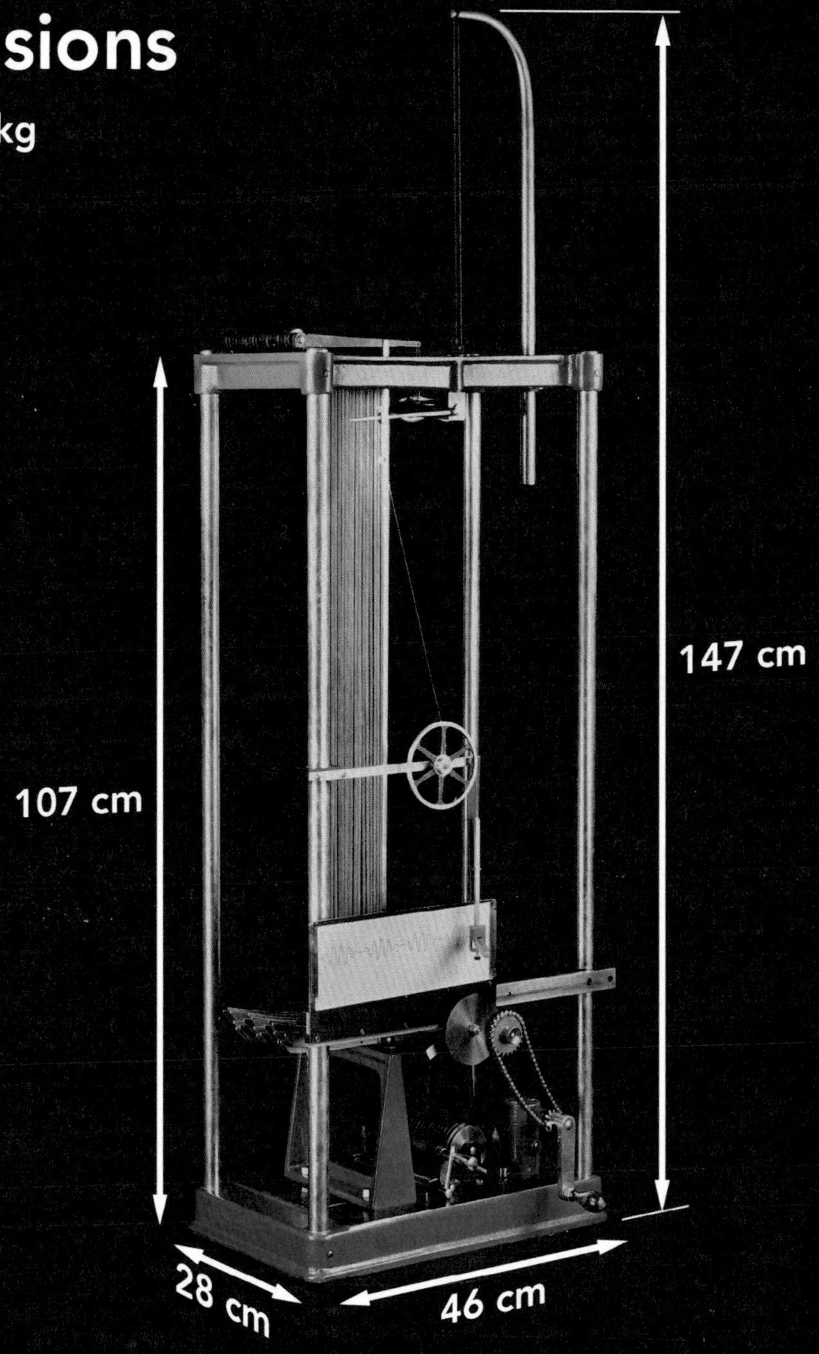

Visual table of contents

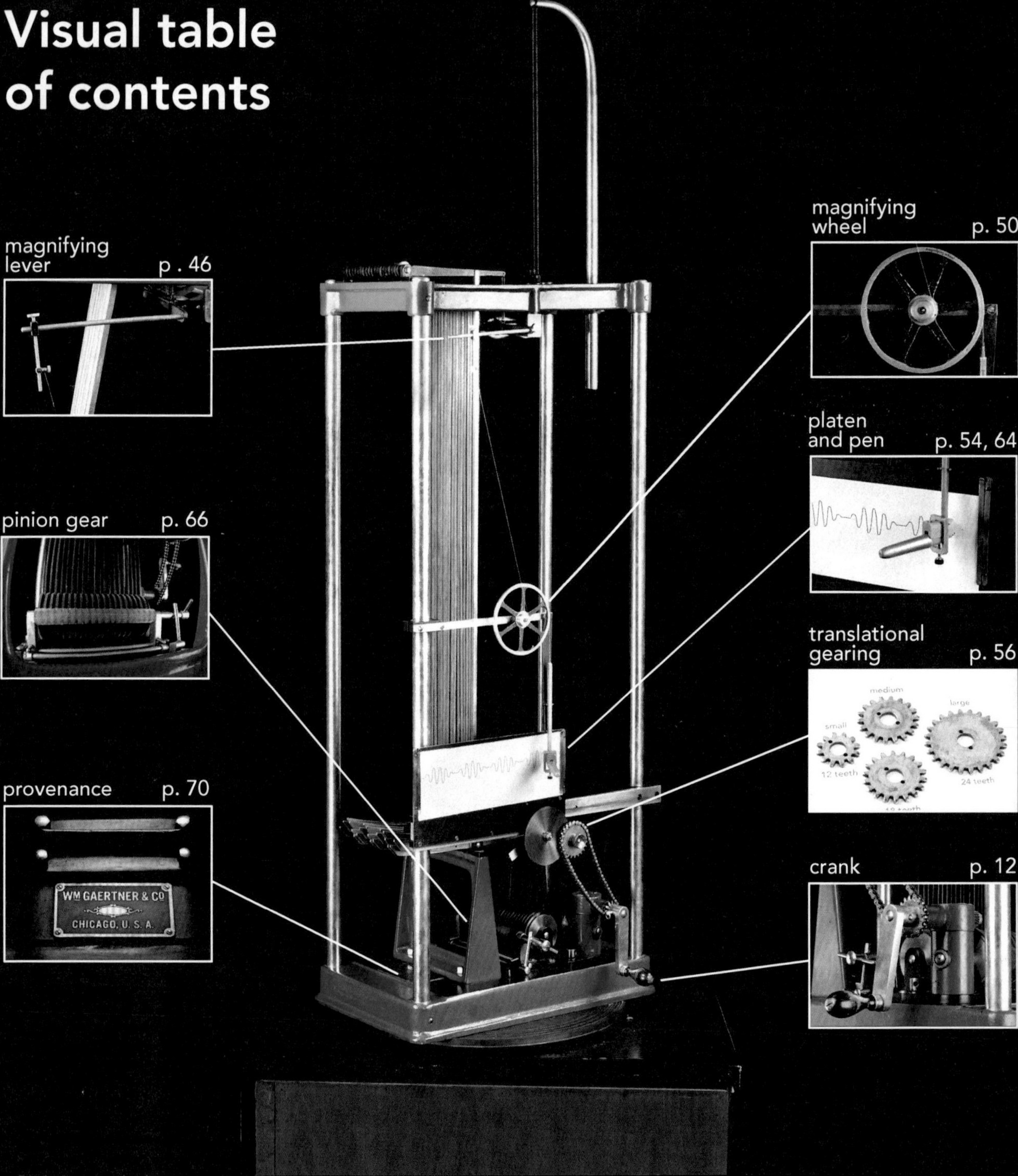

magnifying lever p. 46
pinion gear p. 66
provenance p. 70
magnifying wheel p. 50
platen and pen p. 54, 64
translational gearing p. 56
crank p. 12

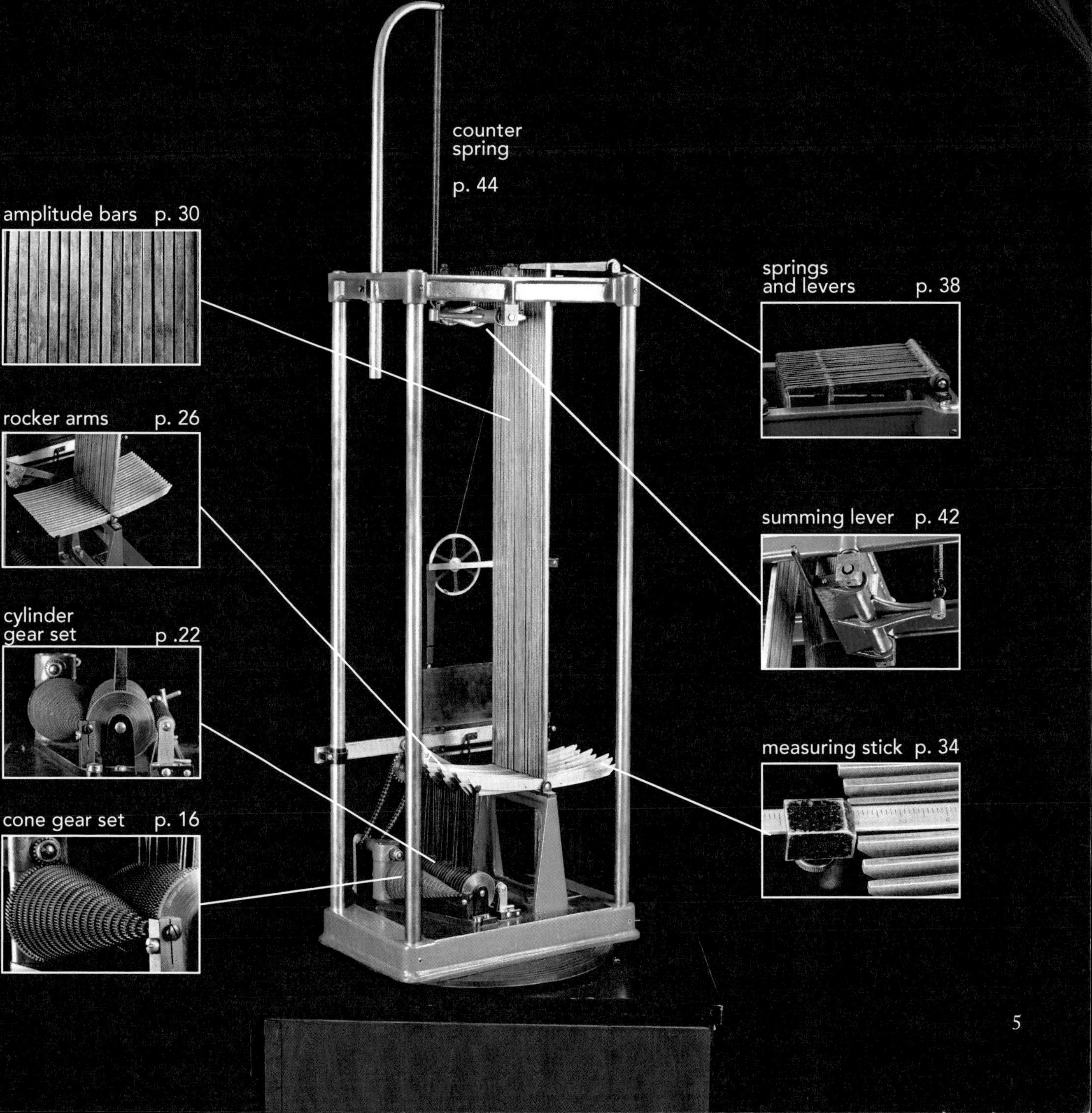
counter
spring
p. 44
amplitude bars p. 30
rocker arms p. 26
cylinder
gear set p .22
cone gear set p. 16
springs
and levers p. 38
summing lever p. 42
measuring stick p. 34

Fourier Synthesis

This machine implements a simplified equation used in Fourier synthesis:

$$f(x) = \sum_{n=1}^{20} a_n \cos(nx)$$

We'll take a look at this equation and then run through it piece by piece to better understand the meaning of each part. On the facing page, we show how the components of the equation are implemented by the analyzer.

$$f(x) = \sum_{n=1}^{20} a_n \cos(nx)$$

A cosine is a wave. It is periodic, which means it repeats after a given period. It always has a value between -1 and 1.

$$f(x) = \sum_{n=1}^{20} a_n \cos(nx)$$

The variable *x* is the position along the horizontal axis of a plot of the cosine.

$$f(x) = \sum_{n=1}^{20} a_n \cos(nx)$$

The value *n* is an integer that ranges from 1 to 20. It determines the frequency—i.e., the number of oscillations—for each cosine in the equation.

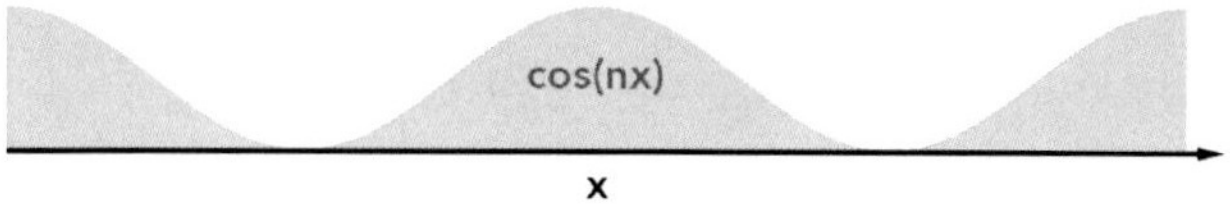

$$f(x) = \sum_{n=1}^{20} a_n \cos(nx)$$

This symbol, a_n (read: "A sub N"), is called the coefficient. The values of a_n (a_1, a_2, a_3, ... , a_{20}) determine the function that will be synthesized.

Each value of a_n is the amplitude of a particular cosine. The value of *n* determines how many oscillations there are in the cosine.

$$f(x) = \sum_{n=1}^{20} a_n \cos(nx)$$

The symbol Σ is the Greek letter sigma. In mathematics it denotes a summation. On the bottom "*n*=1" means that *n* is the index variable which starts at 1. The "20" on the top means that we stop counting *n* once we get to 20.

$$f(x) = \sum_{n=1}^{20} a_n \cos(nx)$$

f(x) is the result of the summation.

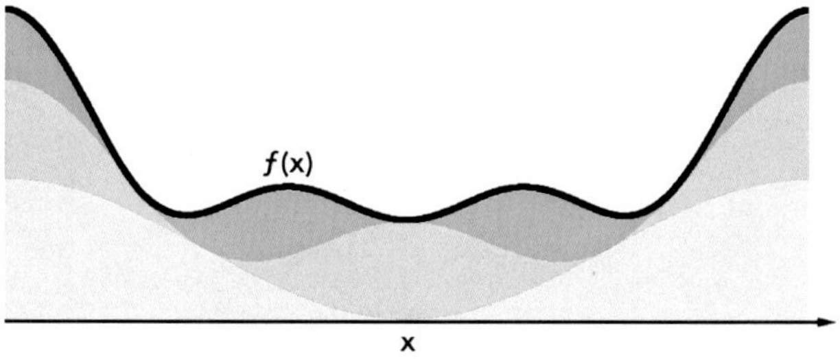

If the cosines are stacked one on top of the other, this adds them together—this is the summation. A line drawn across the top of the stacked cosines is *f*(x), the resultant function.

$$f(x) = \sum_{n=1}^{20} a_n \cos(nx)$$

symbol	mechanism	interpretation	page
x		The variable x is proportional to the rotation of the crank.	12
n		The variable sizes of gears in the cone set drive the gears in the cylinder set at different frequencies. The *n*th gear on the cylinder gear set spins at a rate *n* times as fast as the first gear. There are twenty gears and so there are twenty frequencies produced.	16
$\cos(nx)$	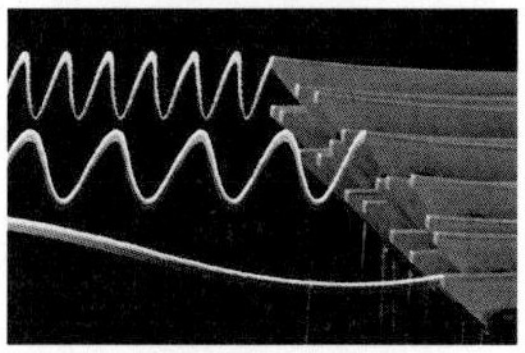	Cams attached to the gears in the cylinder gear set produce near-sinusoidal oscillations at the tips of the rocker arms. Each rocker arm produces its own sinusoidal wave.	26
a_n		The positions of these bars along the rocker arms set the values of the coefficients a_n that weight the sinusoids—there is one bar for each of the 20 frequencies.	30
$\sum_{n=1}^{20}$		A summing lever at the top of the machine adds together the weighted sinusoids.	42
$f(x)$	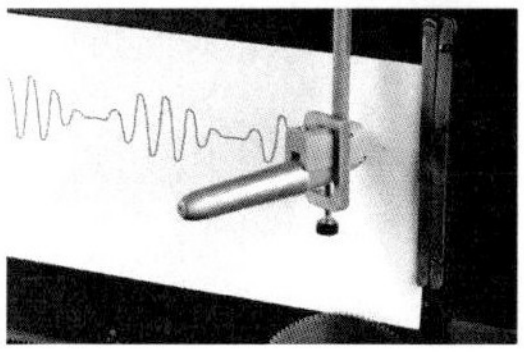	The writing apparatus at the front of the machine plots the resulting sum as a continuous function.	64

Fourier Analysis

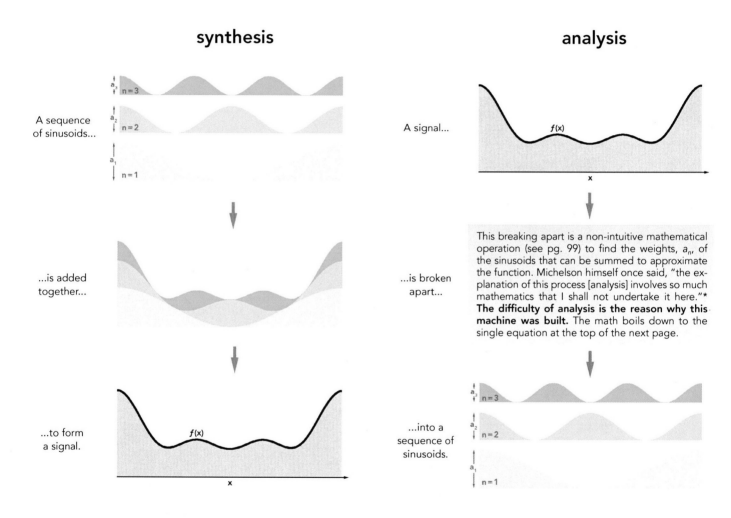

synthesis

A sequence of sinusoids...

...is added together...

...to form a signal.

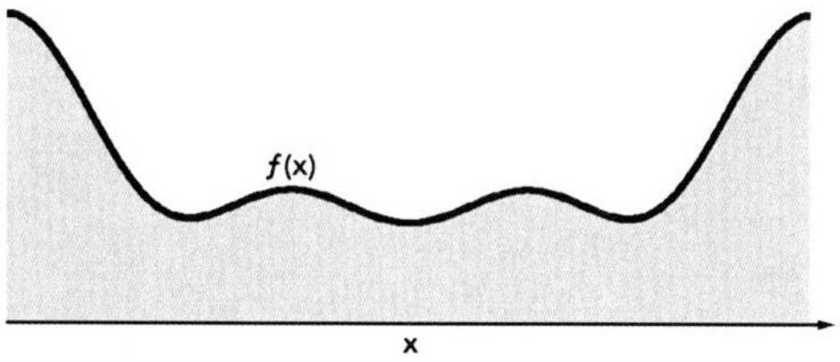

analysis

A signal...

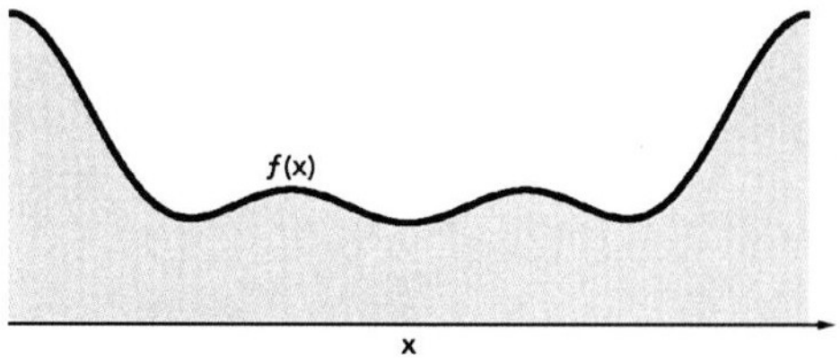

...is broken apart...

This breaking apart is a non-intuitive mathematical operation (see pg. 99) to find the weights, a_n, of the sinusoids that can be summed to approximate the function. Michelson himself once said, "the explanation of this process [analysis] involves so much mathematics that I shall not undertake it here."* **The difficulty of analysis is the reason why this machine was built.** The math boils down to the single equation at the top of the next page.

...into a sequence of sinusoids.

*Albert Michelson, *Light Waves and Their Uses* (Chicago: The University of Chicago Press, 1907) p. 73. Chapter 4 of this book has an excellent discussion of how Michelson used the analyzer in his spectroscopic investigations.

$$a_n \approx \sum_{k=1}^{20} f_k \cos\left(kn\tfrac{\pi}{20}\right)$$

symbol	mechanism	interpretation	page
n		The value of *n* corresponds to the turning of the crank, and more precisely, to the angular rotation of the first gear in the cylinder gear set. Every two turns of the crank increases the variable *n* by 1.	12
k		Each gear in the cylindrical gear set spins with an angular velocity proportional to the size of the cone gear it engages. There are twenty gears and so twenty frequencies.	16
$\cos\left(kn\tfrac{\pi}{20}\right)$	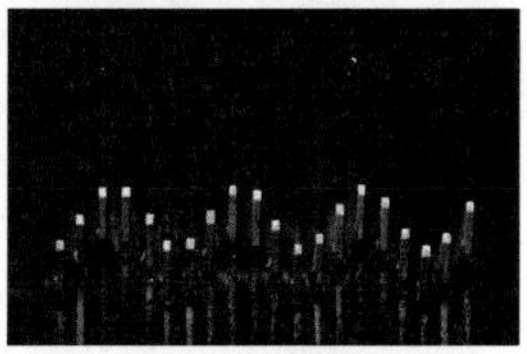	As the crank is turned, sinusoids of varying frequency appear when viewing the tips of the rocker arms from the side of the machine.	26
f_k		The function to be analyzed is sampled at discrete points and these values are used to set the location of the corresponding amplitude bars on the rocker arms.	30
$\sum_{k=1}^{20}$	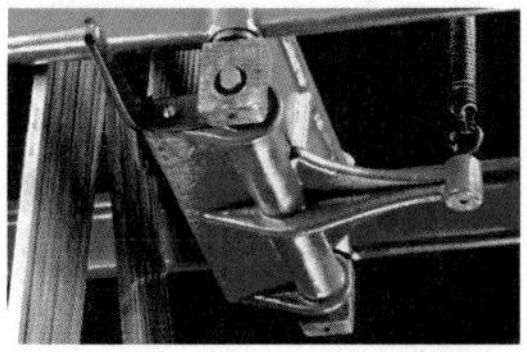	The summing lever at the top of the machine adds together the weighted sinusoids.	42
a_n	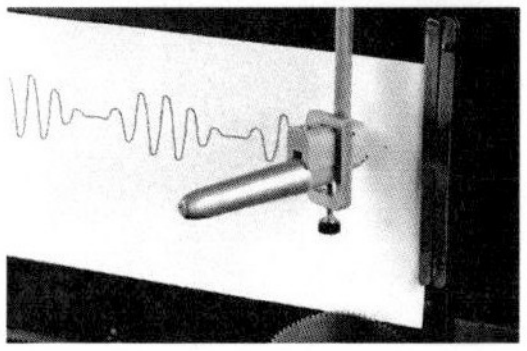	The writing apparatus at the front of the machine plots a continuous function, which is read every two full cranks to yield the approximate value of a_n.	64

The Harmonic Analyzer

Crank

THE CRANK PROVIDES the sole motive power for all the operations of the machine. As the operator turns the crank, the machine comes alive: the gears silently spin, the rocker arms seesaw, the springs elongate and contract, the pen moves up and down, and the paper travels sideways. The handle of the crank, a smooth piece of wood stained black, has a shape well-suited for a firm grip, and it rotates on a pivot so that the operator's hand doesn't slip while turning. Due to the changing positions of the springs, amplitude bars, and rocker arms, the force required to turn the crank can vary markedly as it rotates. A tapered pin, which affixes the crank to a shaft, can be removed so the gear on the crankshaft can be changed (pg. 56). Notice the small fiducial indentations that aid alignment when replacing the crank. The small eyelet on the side of the crank arm once held a small chain, now long lost, that tethered the pin to the crank.

tapered pin
crank

fiducial mark

crank
tapered pin

Cone Gear Set

THE CRANK ROTATES A conically-shaped set of gears, reduced in 4:1 ratio, so that one turn of the crank turns the cone gear set a quarter of a revolution. This cone gear set, not commonly seen in other machines, changes the continuous motion of the crank into the twenty different frequencies needed by the machine. The set consists of twenty different spur gears that are fixed to the same shaft so that they rotate together. Each gear on the cone gear set engages a corresponding gear on the cylinder set at an oblique angle; this lack of full engagement has left distinct wear patterns on the teeth of all the gears, with the smallest gears of the cone exhibiting the most wear. The smallest spur gear has six teeth, the next larger has 12 teeth, with each succeeding gear having 6 more teeth than the gear before, up to the twentieth gear with 120 teeth. The four smallest gears at the tip of the cone are slightly more yellow in appearance and seem to be made of a different, perhaps harder metal. By loosening a knob, the cone gear set can be pivoted out of engagement so that the cylinder set can be aligned for generating either sines or cosines (pg. 66).

cylinder gear set
cone gear set

cone gear set
amplitude bar
rocker arm
connecting rod
cylinder gear set
pivot

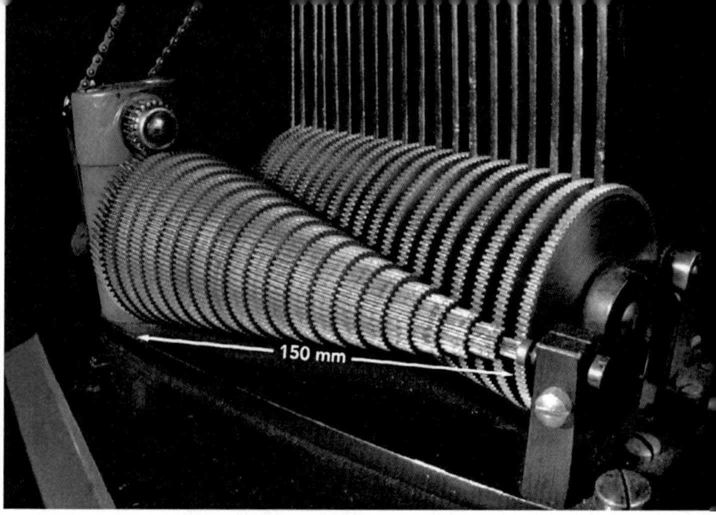
150 mm

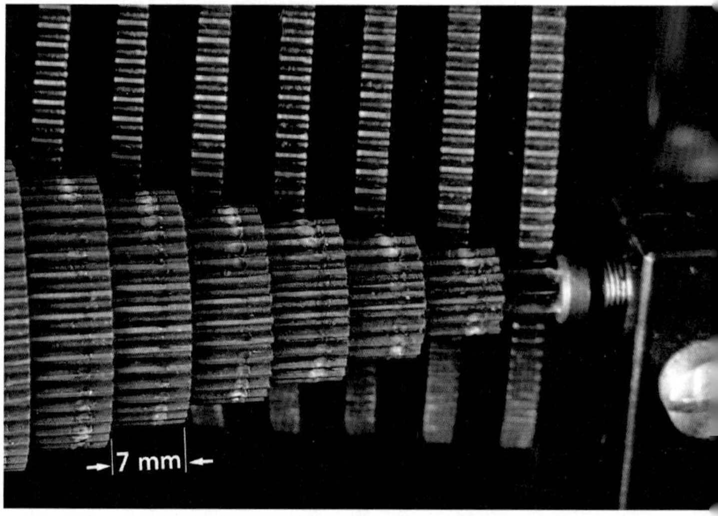
7 mm

This gear engages the crank
7 mm
20th gear 120 teeth
19th gear 114 teeth

Cylinder Gear Set

THE 20 GEARS ON THE cone gear set engage a "cylindrical" set of gears. The gears on the cone set have graduated sizes, but all spin at the same angular velocity. In contrast, the gears on the cylinder set are of equal size, but each gear spins independently with an angular velocity proportional to the size of the corresponding gear on the cone set. The cylinder gear set is actually a sandwich, alternating shiny brass gears with black, rough-finished connecting rods. Each connecting rod rides on a cam (see cam outline on pg. 25) attached to the cylinder gear to its right. As a particular gear on the cylinder turns, its cam drives the corresponding connecting rod in a reciprocating up-and-down motion, producing a near-sinusoidal oscillation on a rocker arm attached to the other end of the rod. This combination of mechanical elements produces the twenty different frequencies used in the analyzer. Another feature of the cylinder gear set, one easily overlooked on cursory inspection, is that each gear contains a notch, approximately 3 mm in depth, that is used to align the gears on the cylinder as well as to set up the analyzer to calculate with either sines or cosines (see pg. 66 for a description of the alignment process).

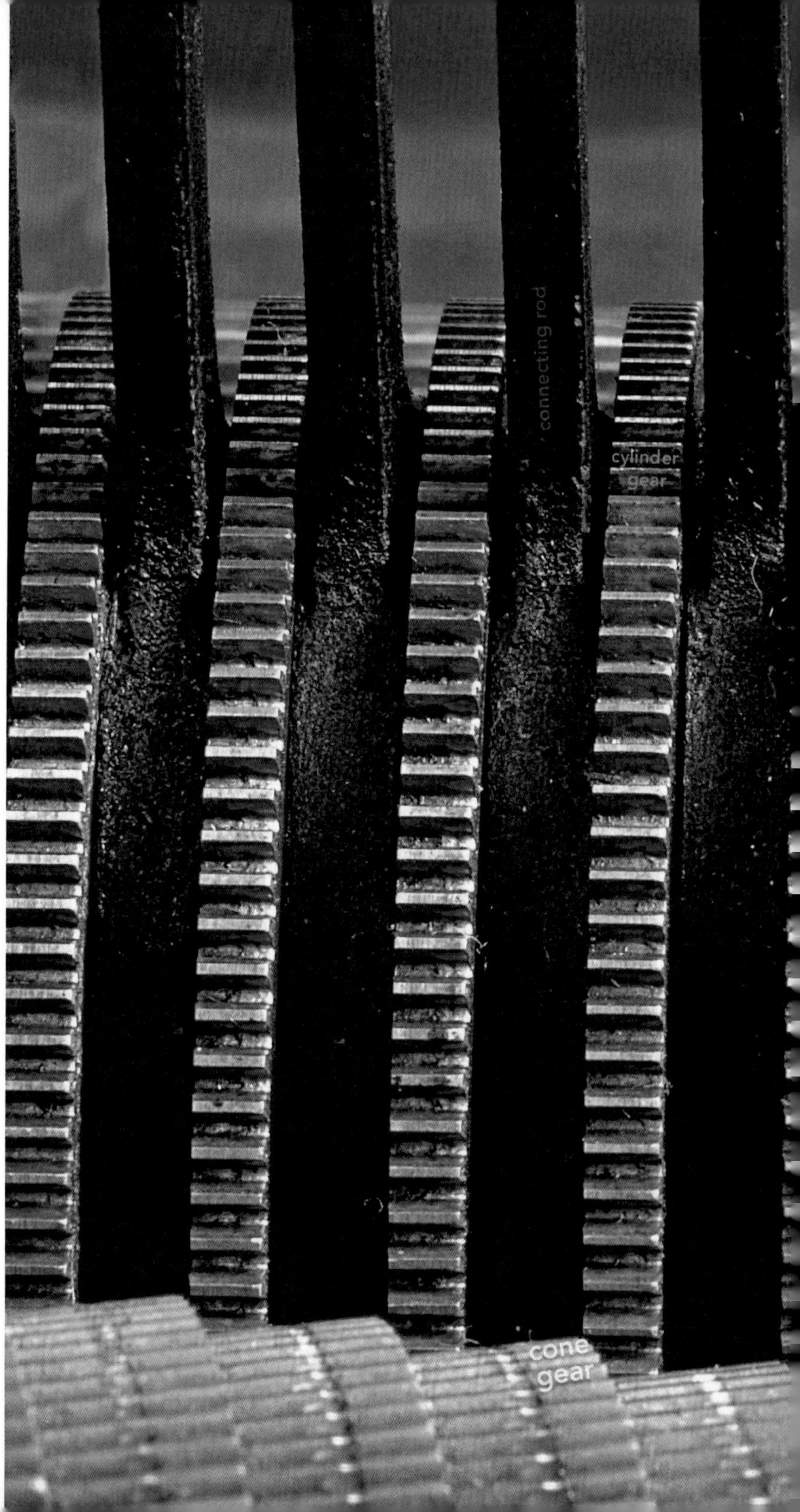

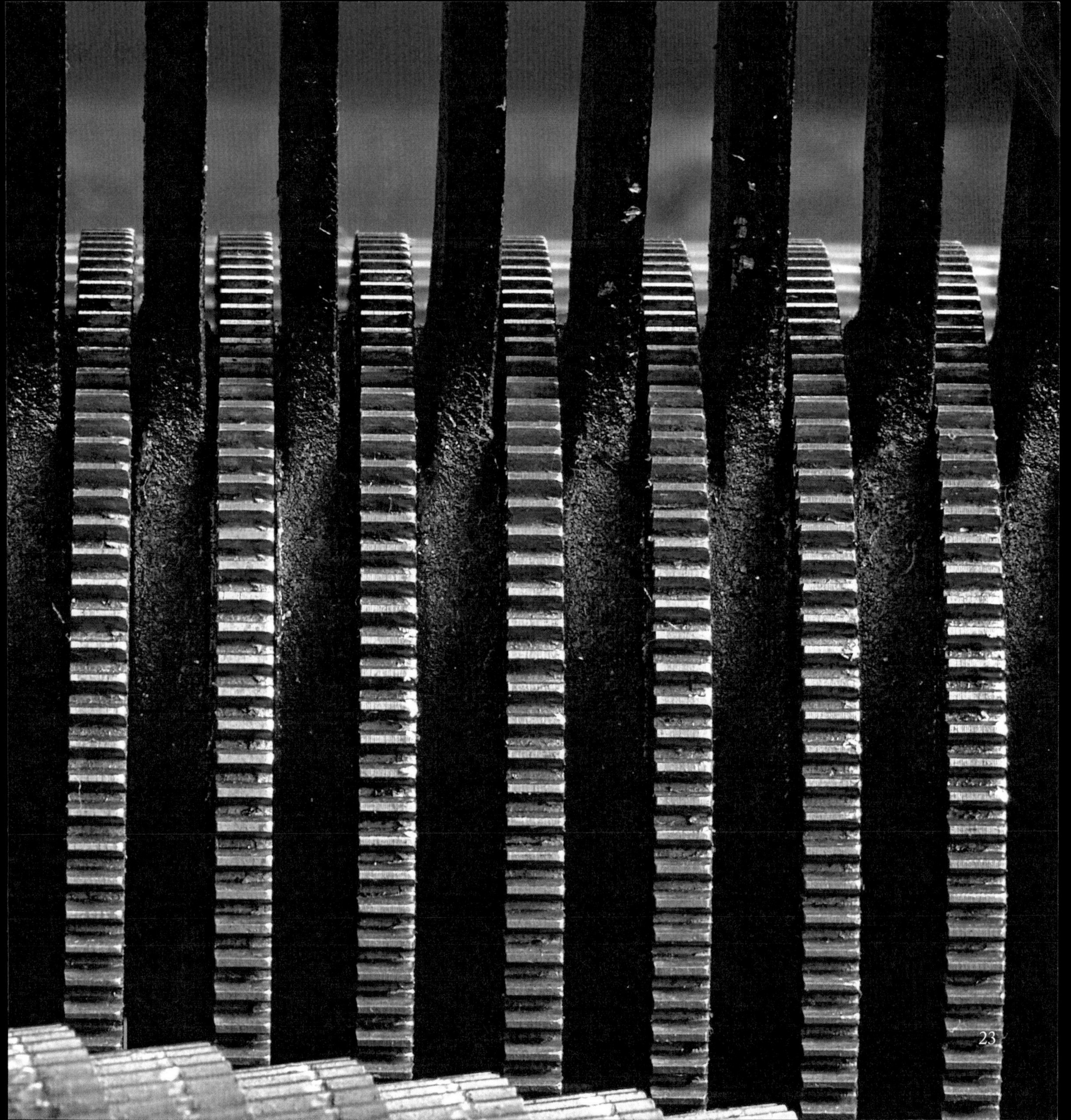

front
side

notch

notch
back side

connecting rod
cam

connecting rod

Rocker Arms

A SET OF MATTE-BLACK vertical connecting rods transfer the oscillatory motions of the cams associated with the cylinder gears to a set of rocker arms. The arms are shaped concave upwards with a radius of curvature the same as the length of the amplitude bars that ride on them. As the crank is turned, the motion of the ends of the rockers is fascinating to watch (see the pictures on the left of the following spread): each individual rocker arm seesaws up-and-down in a continuous near-sinusoidal motion, at a frequency determined by its corresponding cylinder gear. And at the same time, when viewed from the side of the machine, the ends of the rocker arms show a mesmerizing collective motion: the ends are discretized samples of a sinusoid with frequency determined by the total number of crank turns.

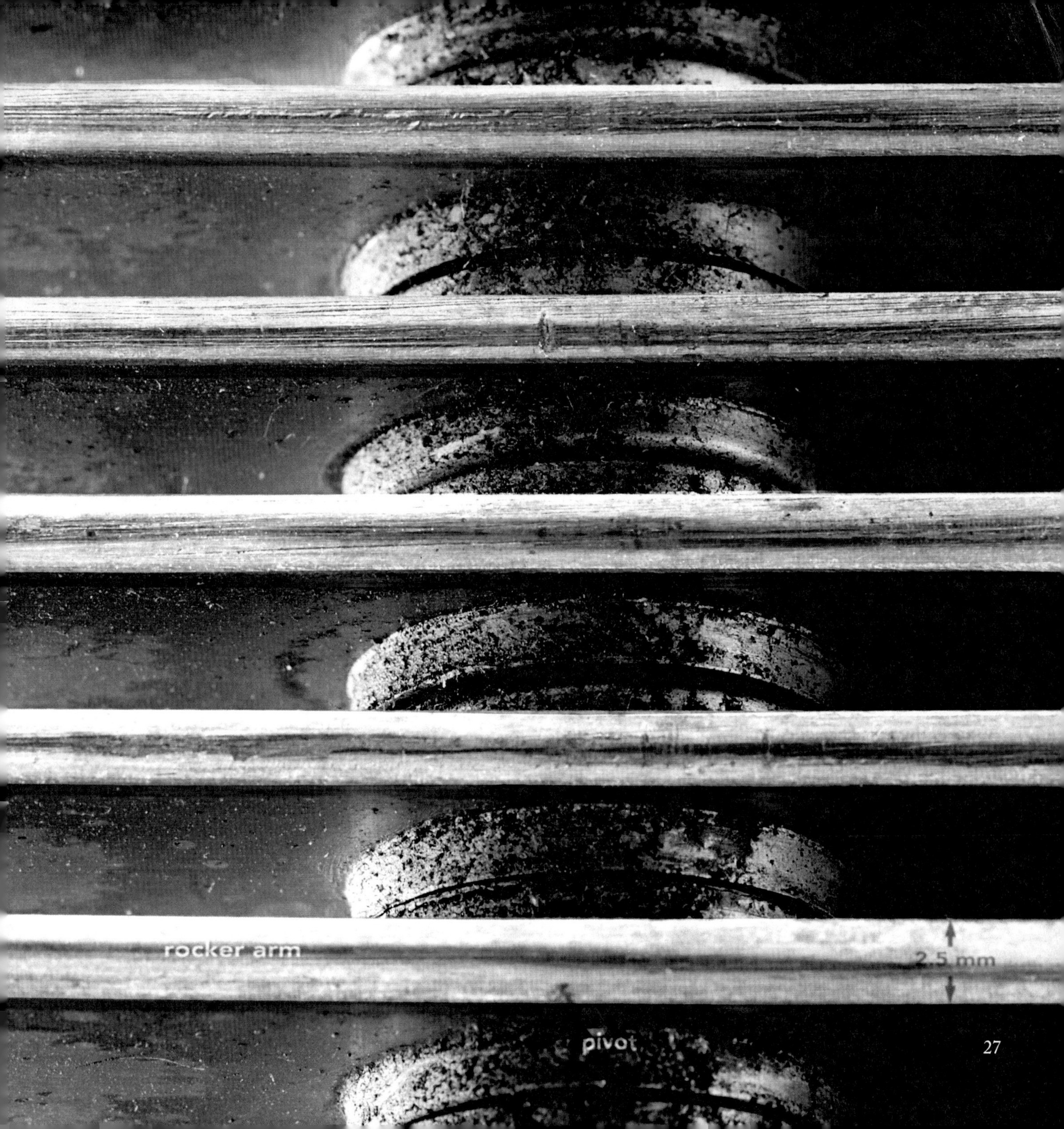
rocker arm
2.5 mm
pivot

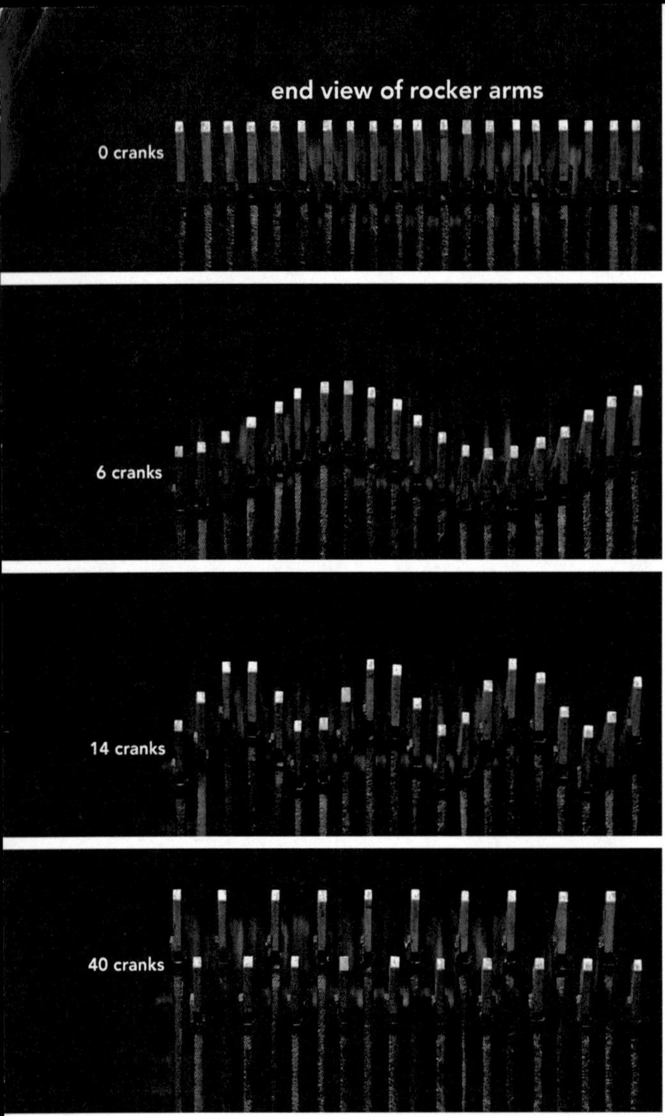
end view of rocker arms
0 cranks
6 cranks
14 cranks
40 cranks

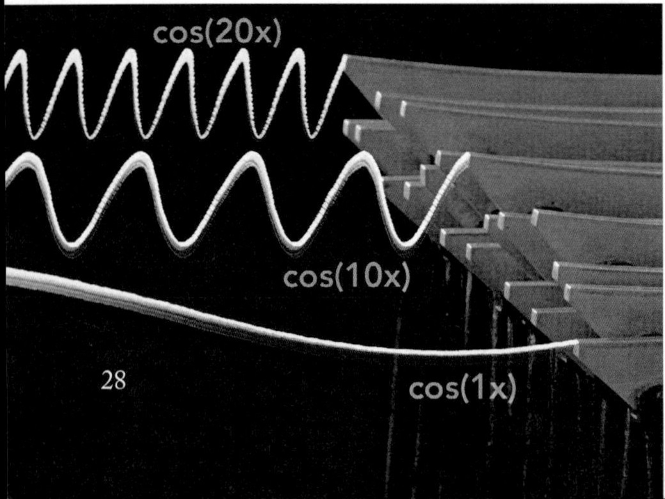
cos(20x)
cos(10x)
cos(1x)

rocker arm
2.5 mm
pivot

16 mm

Amplitude Bars

TWENTY LONG VERTICAL RODS, about 80 cm in length, run up the spine of the analyzer; their chrome-like finish glistens, albeit marred by patches of rust. These rods are called amplitude bars, and their long length ameliorates the nonlinearity inherent in transmitting the weighted sinusoidal motion of the rocker arms to the spring-loaded levers at the top of the analyzer. The position of a particular bar along its rocker arm determines the weighting coefficient, or amplitude, for the corresponding sinusoid. At the bottom of each amplitude bar there is a notch that lets the bar slide along its rocker arm for positioning. While being positioned, a bar produces a satisfying metallic squeak—virtually the only sound the machine ever makes, even during operation. Positive amplitudes are set by positioning the bar on one side of the rocker arm pivot, negative amplitudes on the opposite side. Positioning a bar at the pivot point of its rocker arm sets that coefficient to zero. Care must be taken by the operator during positioning because the bars can slide completely off the rocker.

amplitude bar
rocker arm

6.35 mm
6.35 mm

top lever
amplitude bar

Measuring Stick

To set the amplitude bars on the rocker arms the machine manufacturer, Wm. Gaertner & Co., provided a ruled brass gauge with a stop that slides and locks. The gauge is marked 0 to 10, but the scale is not inches, nor centimeters, but just the 10 equal divisions of one half of the rocker arm. To use it one first sets the value of the coefficient—"2.0" and "9.3" as shown on the pages that follow—lays the stick on the rocker arms, and then slides the amplitude bar, which screeches slightly, out to meet it. Note that the markings are hand stamped, and that the tick mark for 0.5 is longer than any other. Also, some of the markings are unevenly spaced—the distance between 0.4 and 0.5 is smaller than the distance between 0.5 and 0.6—which indicates that the measuring stick was handcrafted. For illustrative settings of the amplitude bars see page 78.

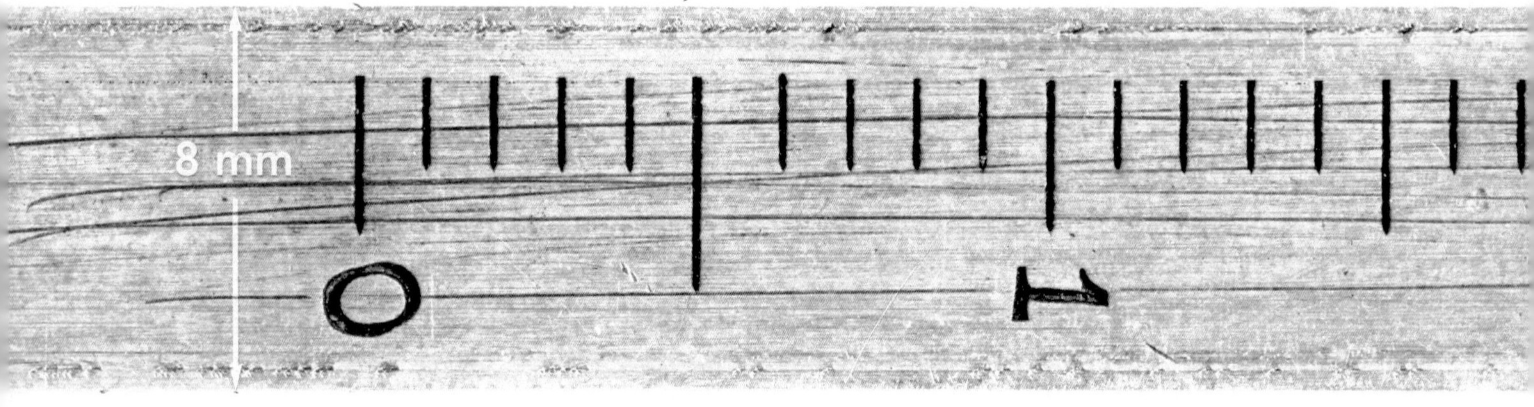

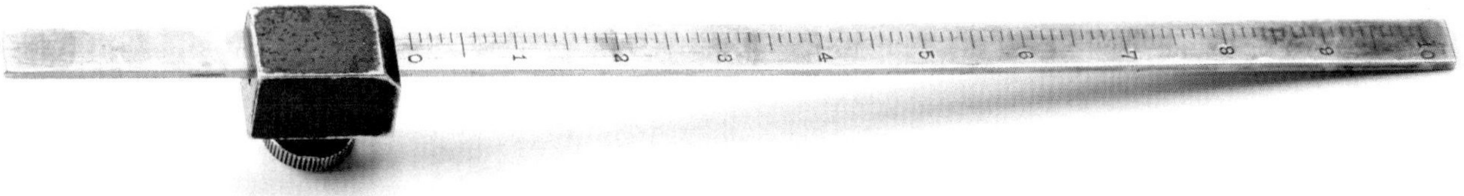

200 mm

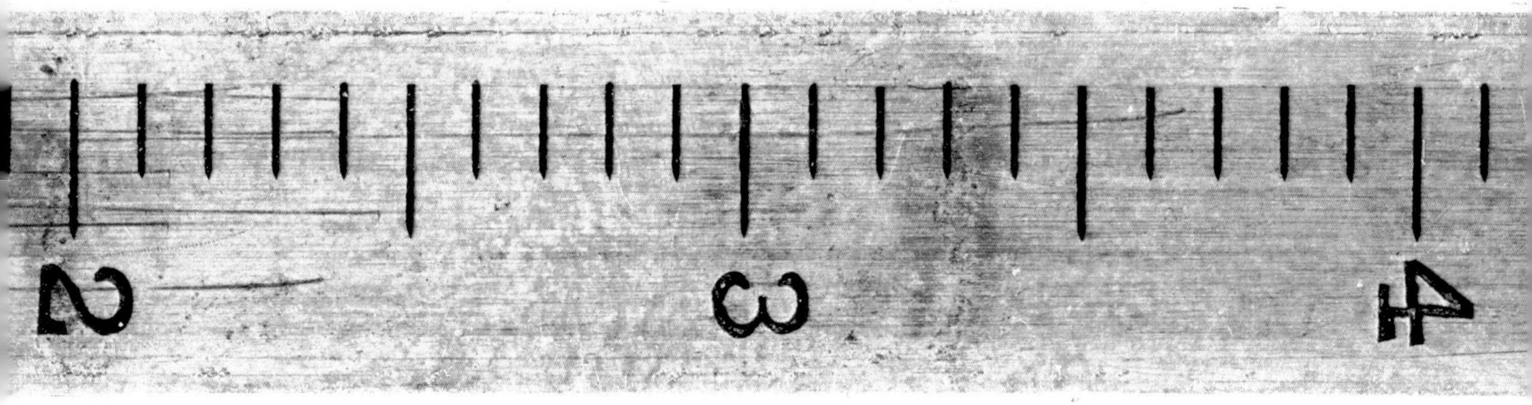
2
3
4

measuring stick
amplitude bar
rocker arm
cylinder gear set

Springs and Levers

AT THE TOP OF THE machine twenty cast metal levers move up and down in sync with the rocker arms. A pivot at each lever's end is the fulcrum, and the load at the lever's opposite end comes from the pull of one of twenty springs attached to a pivoted summing lever. The motion of these third-class levers mirrors that of the rocker arms, but modulated by the positions of the amplitude bars. If an amplitude bar rests in the middle of a rocker arm (at the pivot point) the lever at top stays motionless; if the amplitude bar has been slid to one of the edges of the rocker arm the lever's motion reflects the full amplitude of the tip of the rocker arm; and if the amplitude bar is slid fully to the opposite end of the rocker arm, the lever's motion is 180 degrees out-of-phase, so that when the rocker arm rises, the lever at top drops.

top lever
amplitude bar

top levers
springs

amplitude bar
top lever
springs
summing lever

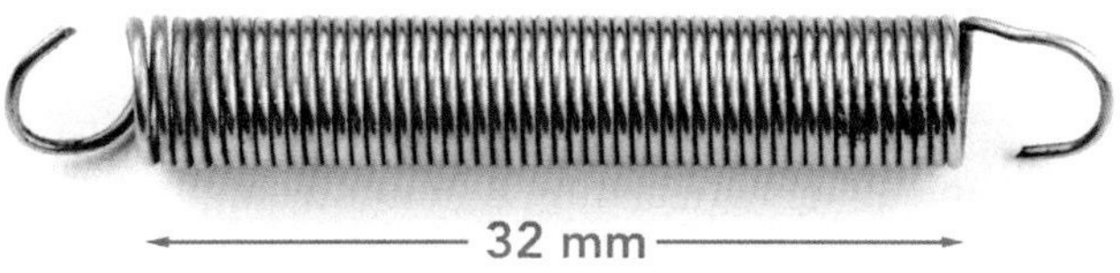
32 mm

Summing Lever

THIS HARMONIC ANALYZER is very tall in relation to its base in order to accommodate the motions created with every turn of the crank. The results of these motions are quietly summed at the top of the machine by an oddly-shaped summing lever. On the end of the summing lever that connects to the twenty small springs from the top levers, it is wide and flat; the other end is long and narrow and connects to a single larger spring which provides counterbalance. The springs on both sides hold this first-class lever in suspension, and its fulcrum is a knife edge in order to reduce friction. The range of motion of the summing lever is very small, on the order of only a few millimeters. The analyzer has mechanisms that bring these motions to human-scale by magnifying and recording them.

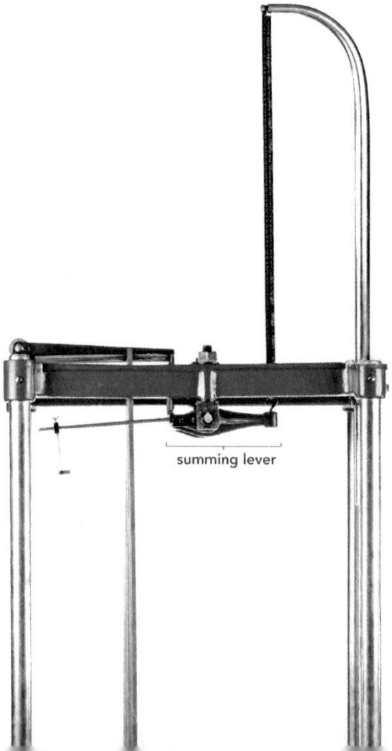

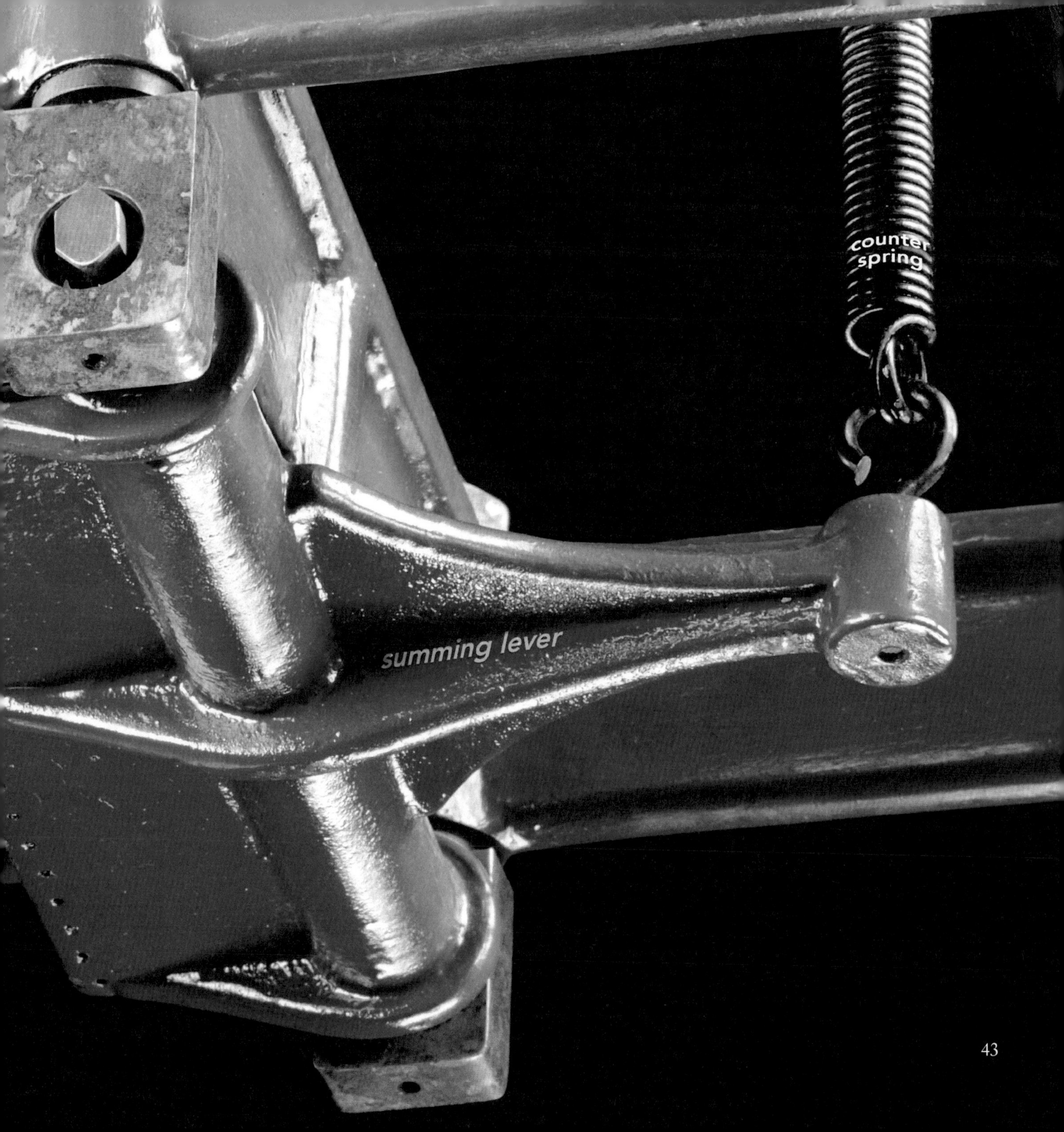
counter spring
summing lever

Counter Spring

A LONG SPRING TOWERS above the machine. One end of the spring connects to a hook on the pivoted summing lever; the other end connects to a curved, tapered post. This large-diameter spring counterbalances the accumulated pull on the summing lever of the twenty individual smaller springs. The machine's operator changes the tension on this counter spring by loosening a square-head screw and adjusting the height of the post up or down. Close examination reveals gouges that mar the finish of the post that were left by the screw during previous height adjustments of the counter spring.

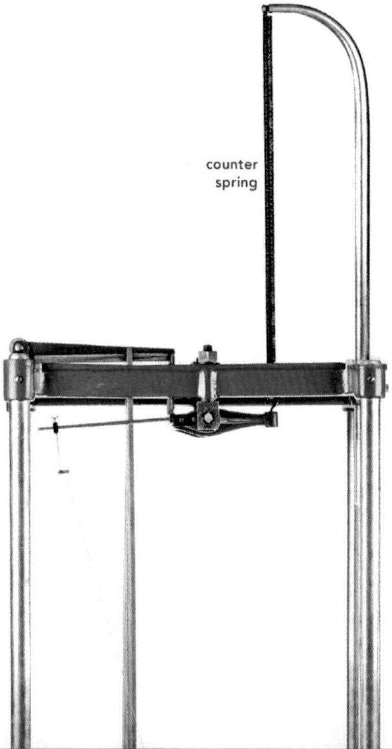

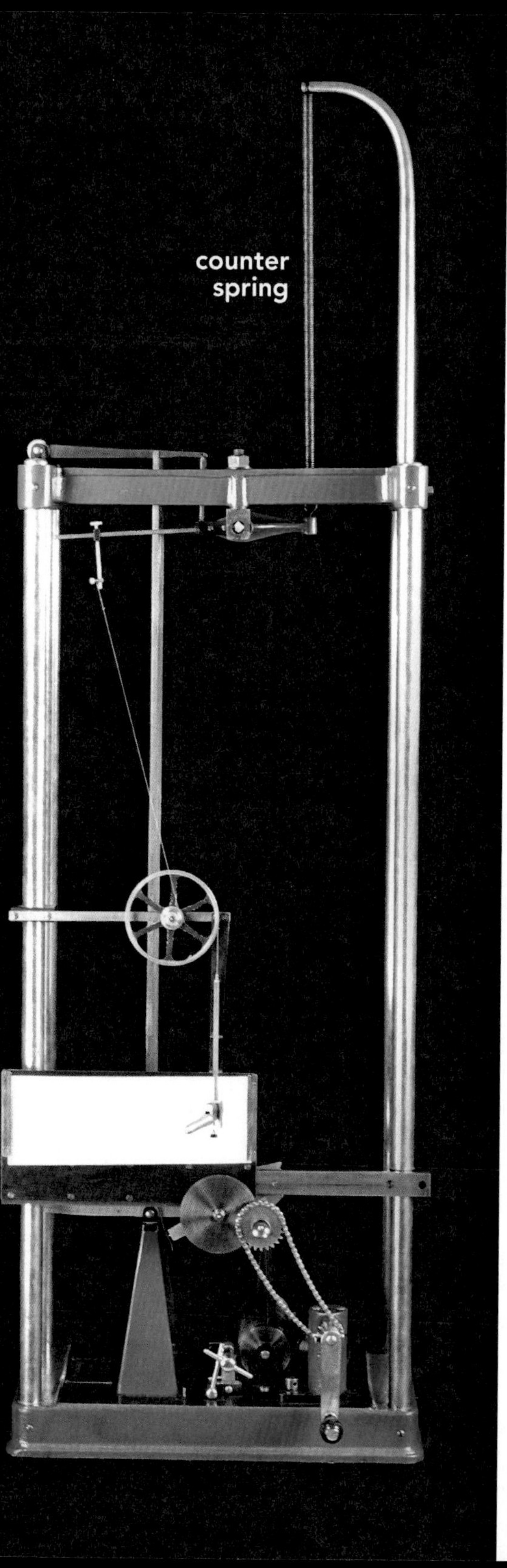
counter
spring

Magnifying Lever

Even though the combined force of twenty small springs tugs at one end of the summing lever, its resulting motion sweeps out an arc of only a few millimeters. This motion must be magnified to produce useful output. Firmly affixed to the summing lever is a round brass rod that magnifies the sweep of the summing lever up to a factor of four. The motion of this rod, called the magnifying lever, is communicated to the writing apparatus below by a long wire attached to a smaller vertical rod. The operator sets the amount of magnification by sliding this vertical rod along the magnifying lever and then tightening a reeded screw to keep it in place. The operator can also adjust the vertical placement of the machine's output by sliding a fixture up and down on the vertical rod. A wire is hooked to this fixture and communicates the motion to the magnifying wheel.

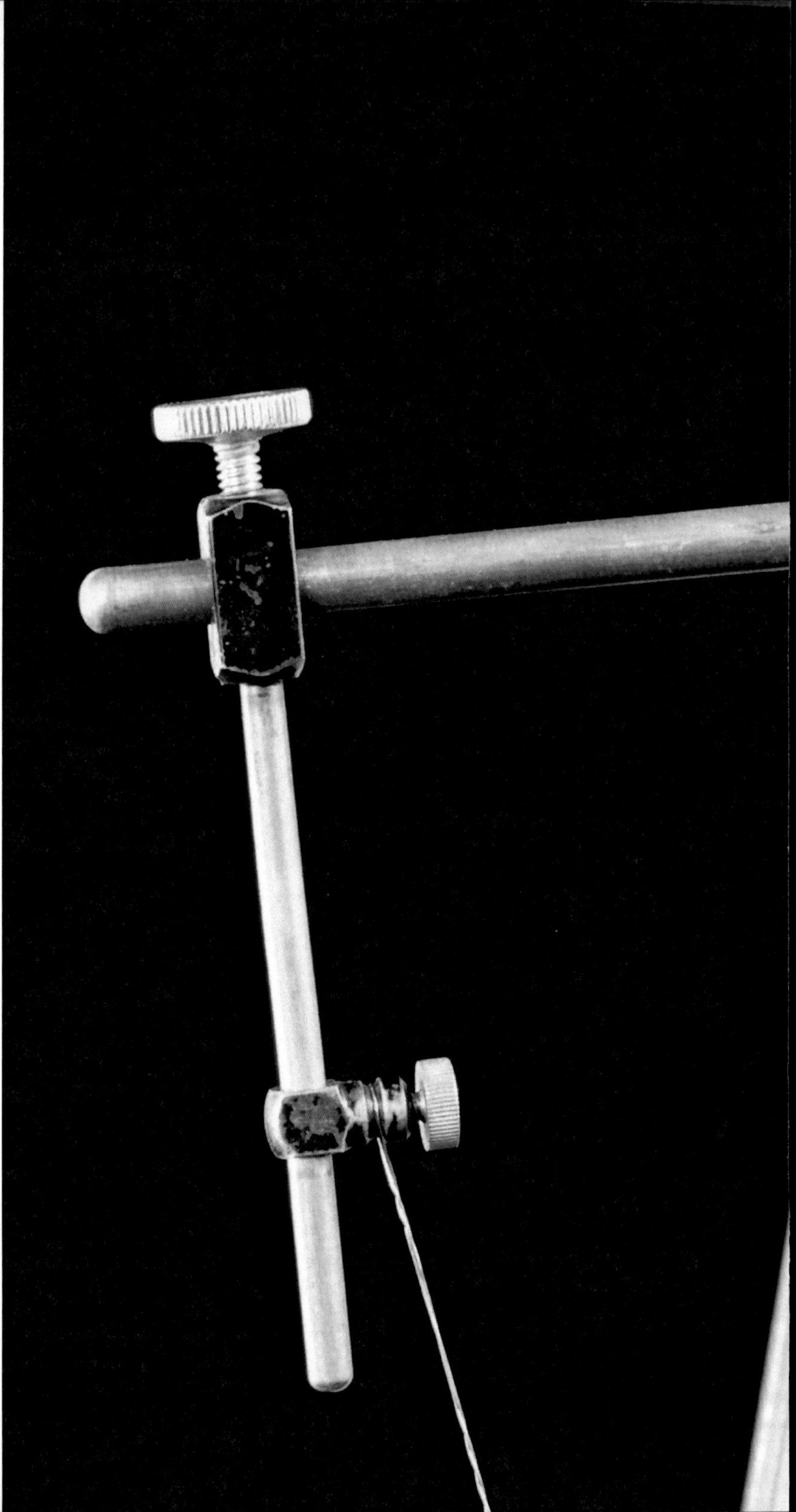

summing lever
magnifying lever
amplitude bars

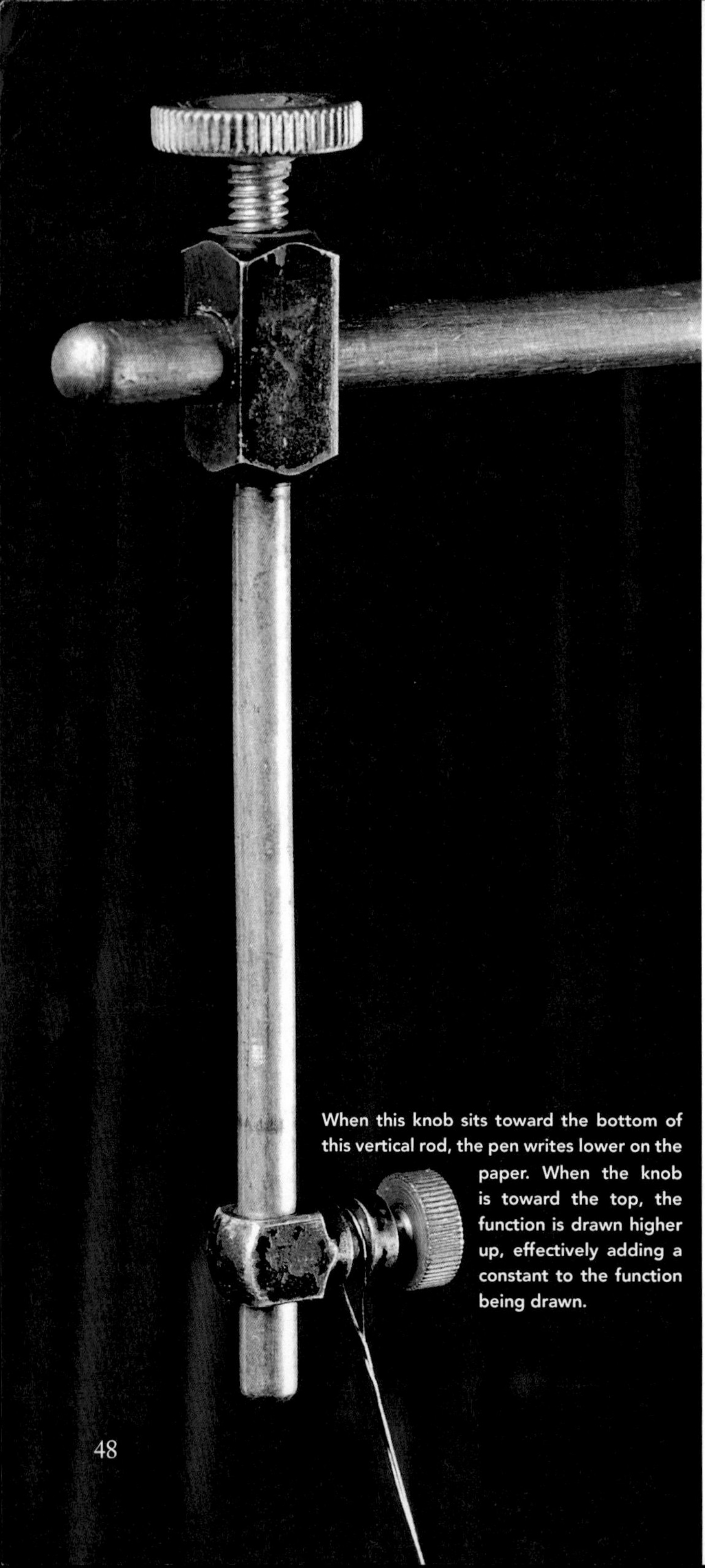

When this knob sits toward the bottom of this vertical rod, the pen writes lower on the paper. When the knob is toward the top, the function is drawn higher up, effectively adding a constant to the function being drawn.

Adjusting the vertical amplification

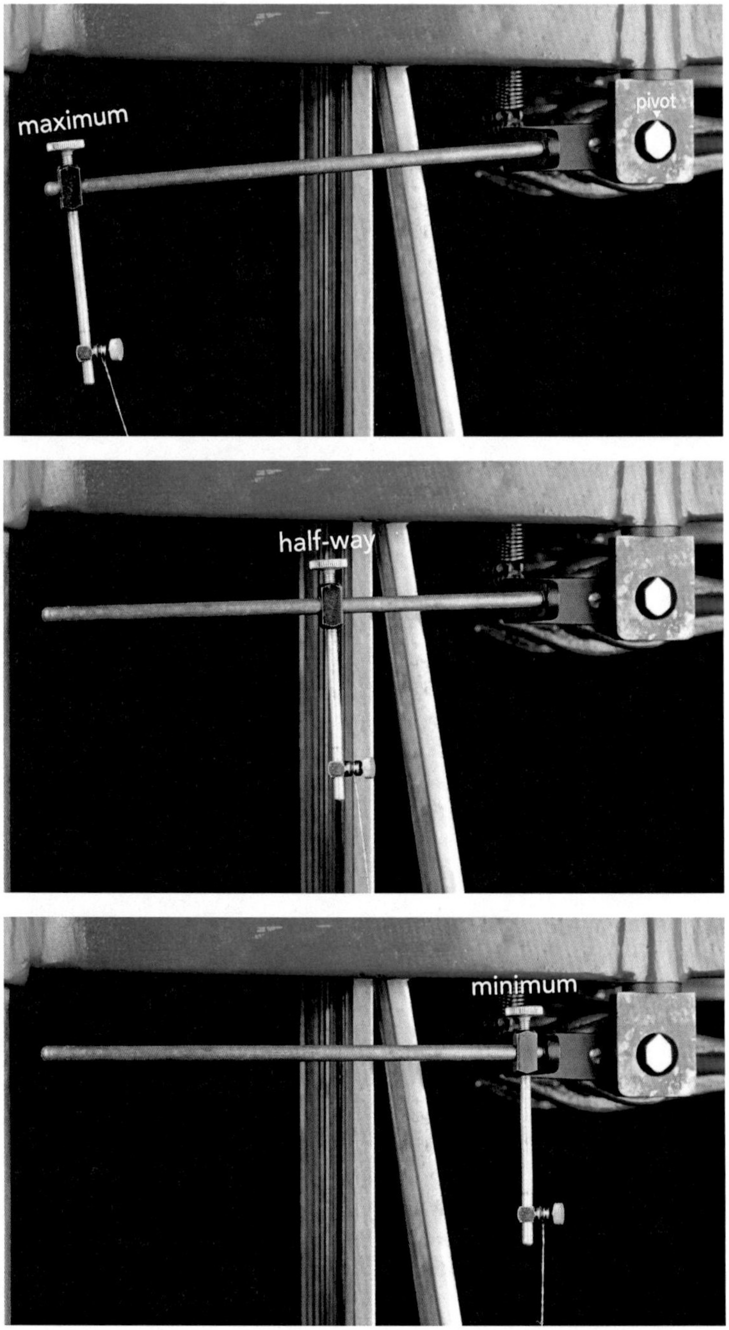

lever setting **output**

maximum

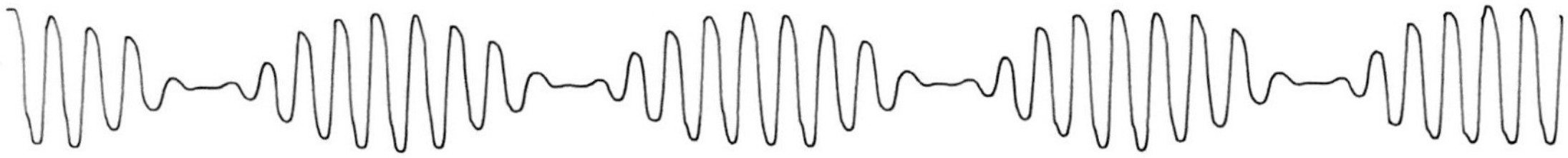

half-way

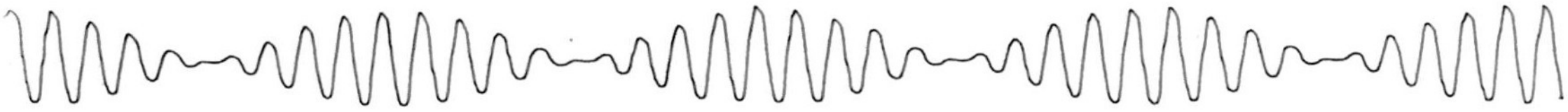

minimum

Magnifying Wheel

A THIN WIRE ATTACHED to the magnifying lever pulls on the inner hub of a magnifying wheel. This magnifying wheel is a pulley comprised of two coaxial wheels that rotate together: a small inner wheel (the hub) and a larger outer wheel. The wheel oscillates as the operator turns the crank; its circular motion mirrors the peaks and valleys of the output. A separate wire is wrapped around the larger wheel and attaches to the pen mechanism. The diameter of the outer wheel is five times the diameter of the inner wheel (100 mm versus 20 mm) and so the motion from the end of the magnifying lever is magnified by a constant factor of five. This wire attaches to the top of the post holding the machine's pen so that the wheel's rocking turns into an up and down motion of the pen. To set up the machine an operator first wraps the outer wire around the hub, holding it in place while looping another wire around the inner hub. If not done carefully the wheel unwinds causing the wires to fly off the hubs and the pen to drop.

to magnifying lever
wire
magnifying wheel
pen mechanism

Platen

THE HEAVY BRASS PLATEN, likely darkened by some treatment, moves a piece of recording paper horizontally while the pen moves vertically. These motions allow for two-dimensional drawing. A toothed brass rack along the platen's bottom edge engages a set of gears driven by the crank. This set of gears shown in the following pages can be unlatched from the platen's rack so that the platen can be moved freely when resetting the machine between calculations. For every new calculation the operator replaces the recording paper by sliding it under the two brass clips on the left and right sides of the platen.

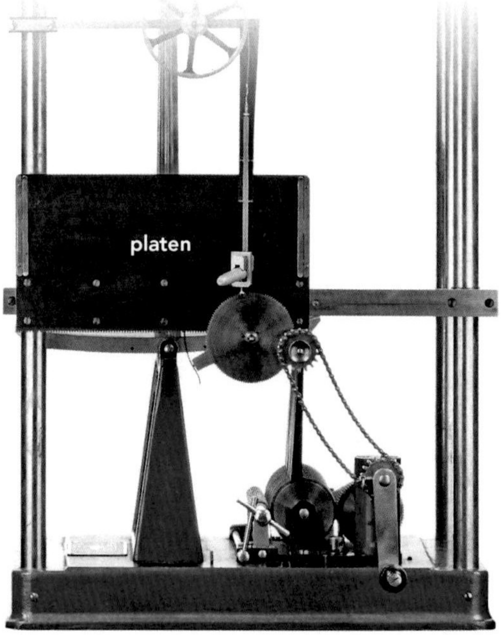

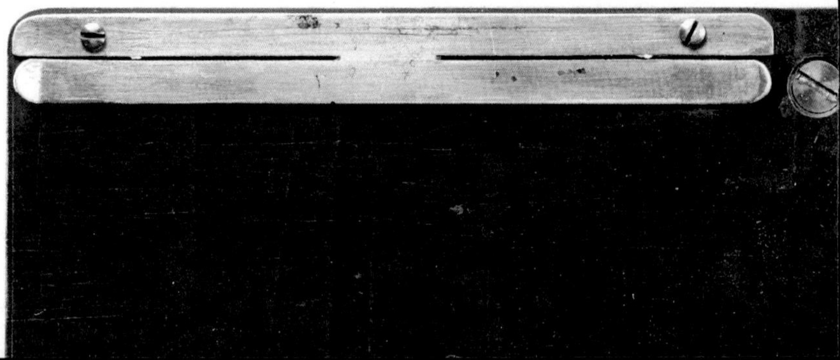

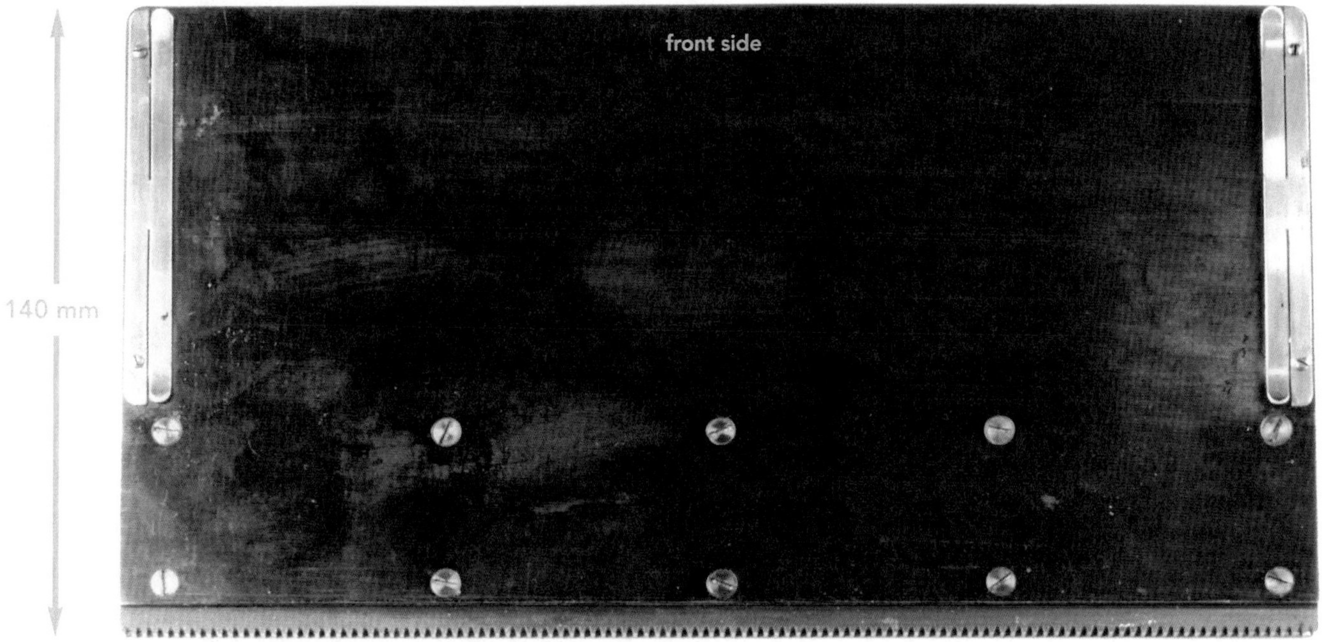
140 mm
front side

back side

Translational Gearing

THE ANALYZER HAS A SET of six translational gears that transfer the motion of the crank to the platen. Because the crank also drives the cone gear set, it must turn through a large number of revolutions to generate the 20 different frequencies. Two of the translational gears are used in a fixed gear reduction of the crank speed. Two of the gears form a rack and pinion that converts the rotary gear motion to linear motion of the platen. The final two gears—the ones connected by a chain, one at the front of the platen, the other on the crankshaft—can be removed and replaced with gears of different sizes so that the operator can fine-tune the speed of the platen as the crank is turned. There is a small latch that allows the operator to disengage the gearing from the platen; this allows the platen to be quickly reset as well as producing slack in the chain for gear replacement. Changing the platen speed changes the horizontal scaling of the output. These two removable gears come in three sizes: small, medium, and large as shown below. Each can be attached to either the platen drive mechanism (upper gear) or the crankshaft (lower gear), as shown on pages 60 and 61. The facing page shows that the small-large gear combination moves the platen the fastest and so yields the greatest horizontal scaling, while the large-small combination moves the platen the slowest and so gives the smallest horizontal scaling.

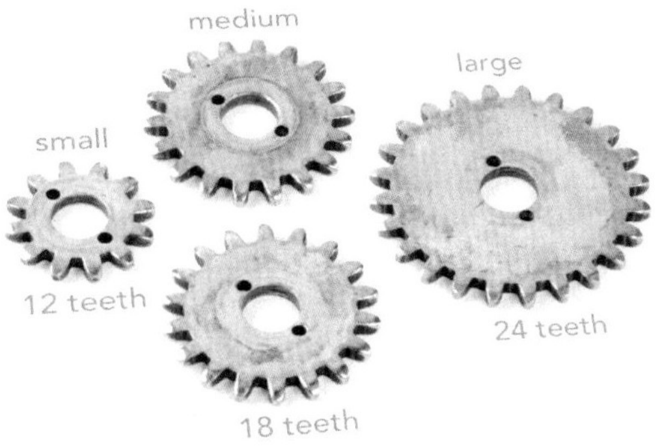

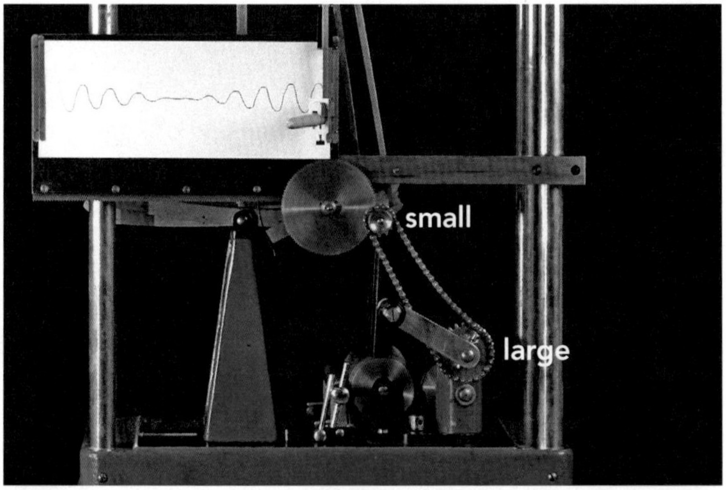

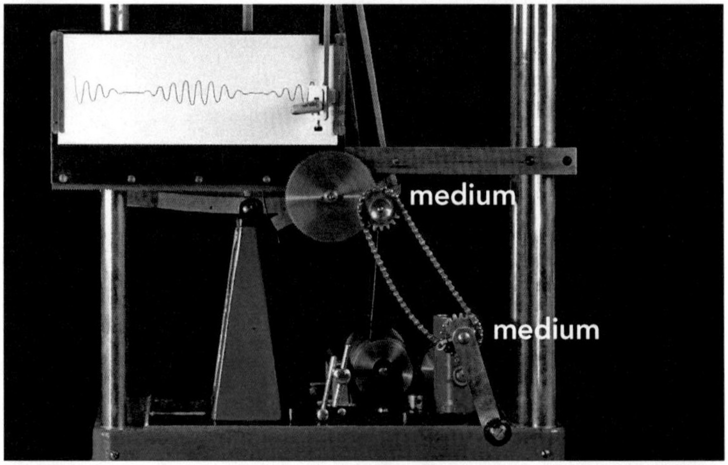

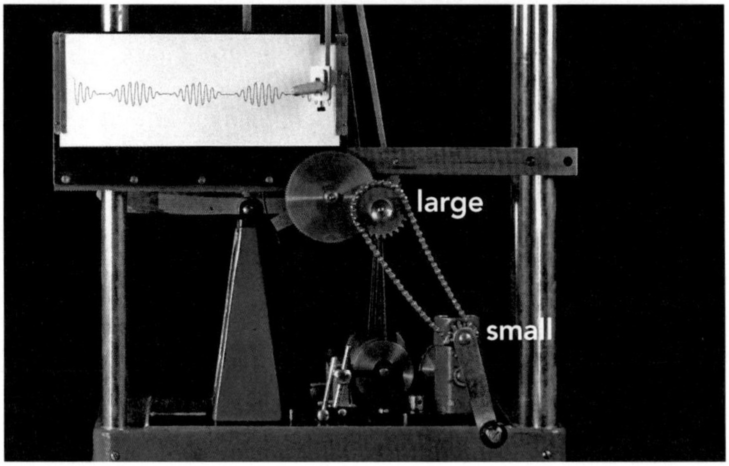

gearing	periods	output
small large	0.5	
medium medium	1	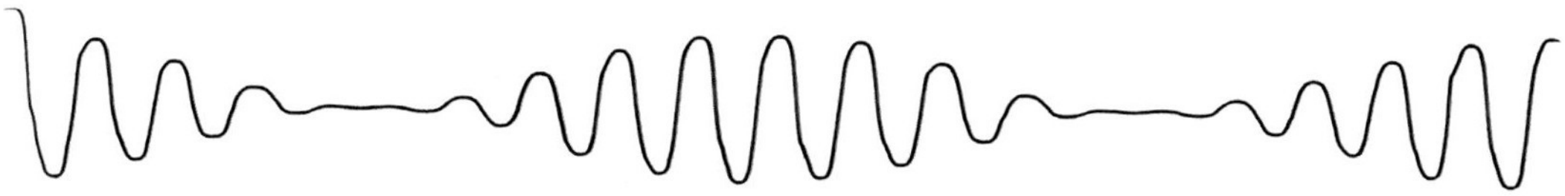
large small	2	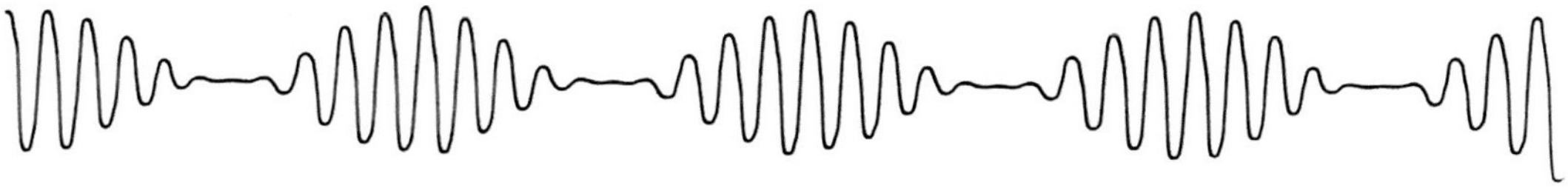

rack

upper gear

lower gear

Looking down, close up on the front of the machine we see the gears that drive the platen. In this shot, the gearing is disengaged from the latch (not visible) and the platen is slid to the left out of frame. The rectangular bar that goes from the top left corner of the image to the right side is the bar that the platen rides on.

Here, we see a view of the translational gearing from the back side of the machine.

Pen Mechanism

THESE BRASS PIECES ARE about 100 years younger than any other part of the machine. This analyzer was missing its writing mechanism for recording the results of its calculations. To rebuild this we reviewed photos and drawings of other Michelson analyzers. In these images, we found several types of writing mechanisms—some machines used a long, horizontal lever arm, like a seismograph, while others had a pen attached to a long, rigid, vertical rod. Ultimately, we built a simple replacement: a brass frame holds a marker in a v-block, which is attached to a square brass rod, which in turn is attached to the wire from the magnifying wheel. The marker, which moves vertically, draws a curve as the platen moves horizontally underneath it. There is also a small set screw that adjusts the angle of the pen to the paper. This allows the operator to reduce the friction between the marker and paper to produce the smoothest output.

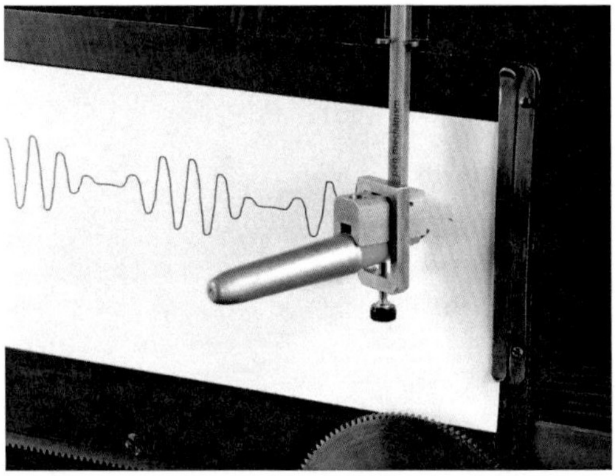

Pinion Gear

THE HARMONIC ANALYZER can calculate using either cosines or sines. Before using the machine, the gears in the cylindrical train must be aligned to ensure that the twenty sinusoids it produces will be in phase at the start. To do this the operator first disengages the cone gear set via a pivot at its tip. Each gear on the cylinder set has a reference mark—a single notch about 3 mm in depth. The operator, using their fingers, rotates each gear in the cylindrical train until the notches line up. After this alignment, a small lever is used to engage the pinion gear with the cylinder gear set. The operator turns the pinion gear, which now moves all the cylinder gears as one, to set the machine to use either sines or cosines. If the notches all point toward the top, the analyzer is set to calculate with cosines; if the notches are 90 degrees from this position, the analyzer calculates with sines. The pinion gear is then disengaged, and the cone gear set re-engaged with the cylinder gear set. Each of these steps tends to move the cylinder gears slightly out of alignment, so constant correction is required.

crank
pinion gear
front
side

cylinder gear set
disengaged
pinion gear
front side

engaged
6 mm

disengaged
gap

engaged

back side

notch

Provenance

THE HARMONIC ANALYZER depicted in this book has a nameplate affixed to its base that declares this particular machine was built by "Wm Gaertner & Co." This small plate, 100 mm by 55 mm, helps date the machine. This company started in 1896 and then changed its name in 1923 to "The Gaertner Scientific Corporation." So this machine must have been built between 1896 and 1923. The manufacturer and the date range for its manufacture are the only solid facts we have about its provenance—to go further requires informed speculation. We don't know who acquired it or even when it arrived at the University of Illinois's Department of Mathematics. The best that we can do is

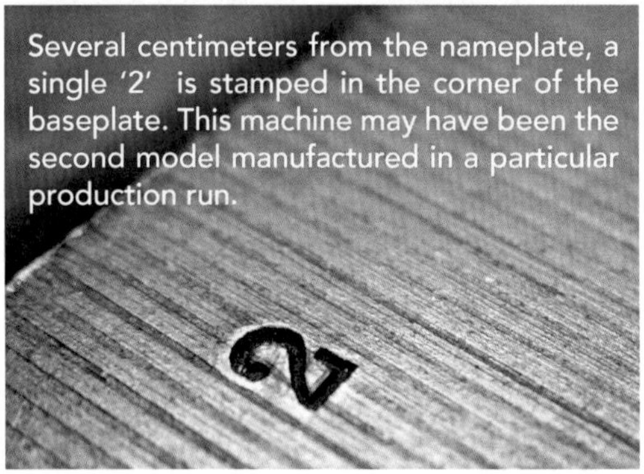

Several centimeters from the nameplate, a single '2' is stamped in the corner of the baseplate. This machine may have been the second model manufactured in a particular production run.

AERTNER & CO
AGO, U. S. A.

report clues and hints based on the machine's design and then correlate those features with information from reports of other Michelson machines built by William Gaertner & Company.

William Gaertner was a German-born instrument maker who worked on the South Side of Chicago until his death in 1948. Gaertner often built commercial versions of the instruments developed by Michelson, then a Professor at the nearby University of Chicago. Gaertner, for example, manufactured and sold the first commercial version of Michelson's interferometer, which was so successful that 50 years after the first one appeared 80% of the interferometers in use in the United States had been built by Gaertner's company.

Gaertner sold harmonic analyzers designed by Michelson in the early decades of the twentieth century. Two versions of the analyzer appear in the company's 1904 catalog, tucked in at the end after pages of interferometers and astronomical instruments. The catalog offers both a 20-element and an 80-element analyzer; it lists no price for either size, but from other research we know that Gaertner did sell some analyzers. The *Columbia University Quarterly* of 1901 highlights the work of "Professor Hallock on the composition of sounds," noting that "he will use a Michelson harmonic analyzer just completed for him by Gaertner, of Chicago." In 1904 the Victoria & Albert Museum reported that "the most valuable acquisition during the year is ... an 80-element Harmonic Analyzer and Integrator, made by Gaertner, of Chicago, to the design of Michelson." The University of Wisconsin's *Biennial Report* for their regents mentions "Details of Disbursements, 1903-04: Wm. Gaertner & Co., harmonic analyzer $412.00." The 1909 sessional papers of Canada—their legislative record—lists "Gaertner, Wm. & Co.: 20 element harmonic analyzer $225." And Ingersoll and Zobel in their 1913

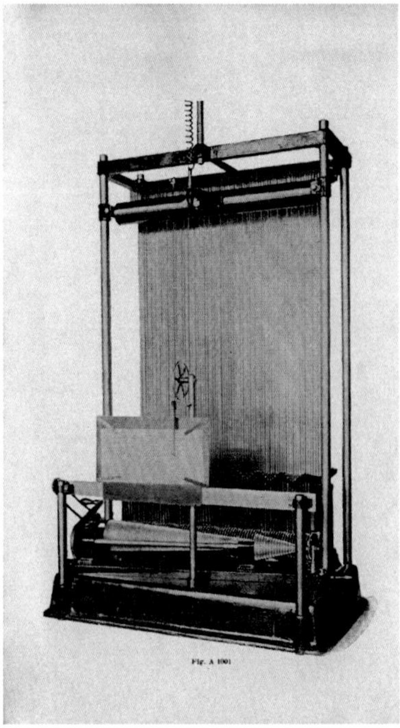
Fig. A 1001

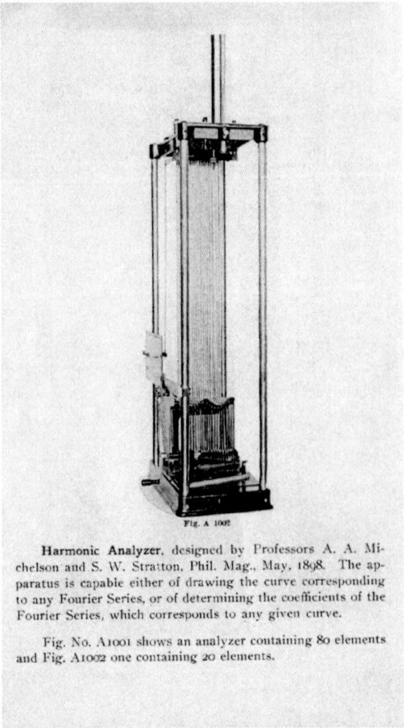
Fig. A 1002

Harmonic Analyzer, designed by Professors A. A. Michelson and S. W. Stratton, Phil. Mag., May, 1898. The apparatus is capable either of drawing the curve corresponding to any Fourier Series, or of determining the coefficients of the Fourier Series, which corresponds to any given curve.

Fig. No. A1001 shows an analyzer containing 80 elements and Fig. A1002 one containing 20 elements.

Left an 80-element machine; *right* a 20-element machine, nearly identical to the analyzer described in this book. The two pages reproduced above appeared in a 1904 Wm. Gaertner & Co. catalog of astronomical, physical and physiological instruments.

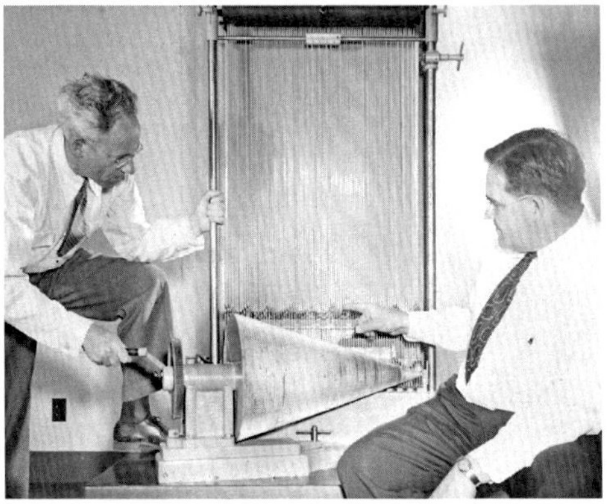

These photographs show Michelson's 80-element analyzer on display ca. 1950-1960. Photos courtesy of the Nimitz Library, United States Naval Academy.

book *An Introduction to the Mathematical Theory of Heat Conduction* describes Michelson's 80-element analyzer and notes "a number of analyzers of this type have been made by Wm. Gaertner & Co. of Chicago." After 1913 we could find no reports of the analyzer until 1933.

At the 1933 World's Fair a 20-element machine was featured in the Great Hall of Science under the title of "The Magic of Analysis." The machine displayed at the fair differed significantly from the analyzer described in this book. Frederick Collins, a British electrical engineer, noted that it was "greatly improved since [the] 1898 machine", specifically "instead of a cone of gears that was used in the first machine, a set of sine cams is used to give motion to the lever arms and tension of the springs." This change in the gear train is confirmed by the recollection of the curator of Mathematics and Antique Instruments at the Smithsonian Institution; in a 1969 interview she recalled that Gaertner still made the analyzer in 1930, but noted that "they changed the design from the cone to the cylinder, and they made some modifications."

So our best guess about this machine's origin and date is that it was one of several 20-element machines manufactured by Wm. Gaertner & Co. between 1896 and 1923 with a high probability that it was made between 1901 to 1909—the era when we see the most reports of 20-element machines. We believe it was purchased for a research project, but, based on the overall lack of wear of the analyzer's moving parts, it was likely never heavily used. The machine now sits proudly in a glass display case in Altgeld Hall, at the University of Illinois at Urbana-Champaign.

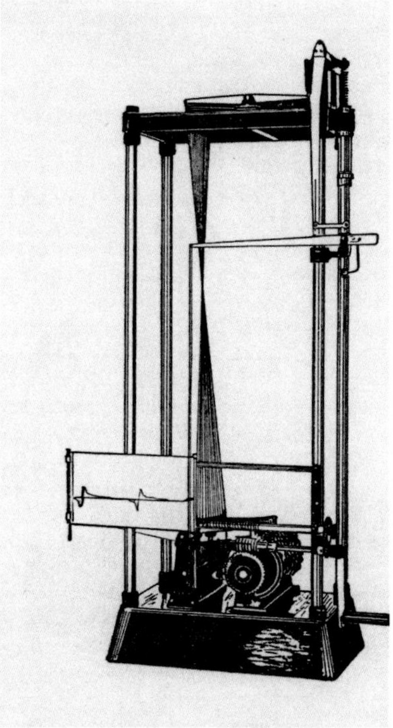

This version of Michelson's 20-element harmonic analyzer appeared at the 1933 World's Fair. Its principal difference from the machine described in this book is the replacement of the cone-cylinder gear set with a set of sine cams.

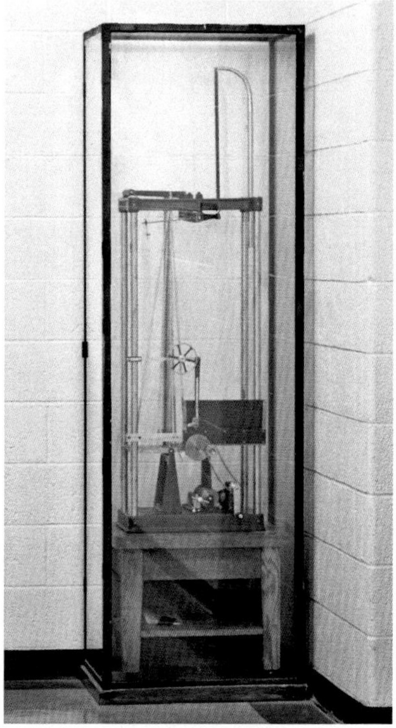

The harmonic analyzer explored in this book rests again where we originally found it: in a glass case in the Department of Mathematics at the University of Illinois at Urbana-Champaign.

Output from the machine

The next fourteen pages show the machine's output for specific settings of the amplitude bars on the rocker arms. In generating this output the machine was set to use cosines, except for the results on page 89 where sines were used.

Pages	Description
76–77	Cosines for all of the twenty frequencies that the machine can produce.
78–81	The amplitude bars are set on the rocker arms to produce four different types of square waves.
82–85	A cosine is sampled at twenty points and placed on the rocker arms. The twenty points span two periods, one full period, half a period and a quarter of a period.
86–88	Arbitrary values are set on the rocker arms.
89	A square wave is set on the rocker arms, but here the machine is set to calculate with sines.

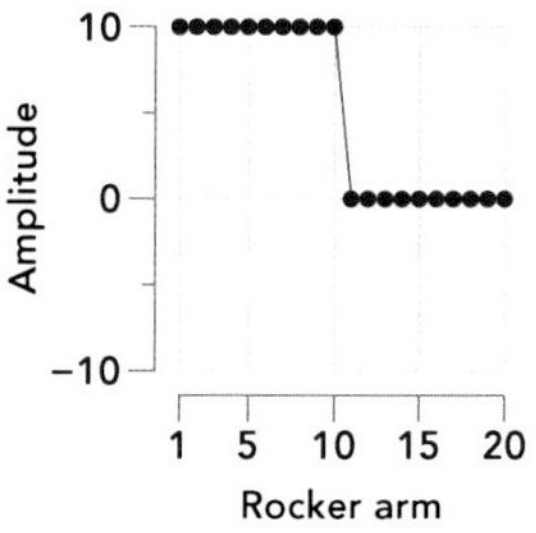

This set of amplitudes ...

... is placed on the rocker arms ...

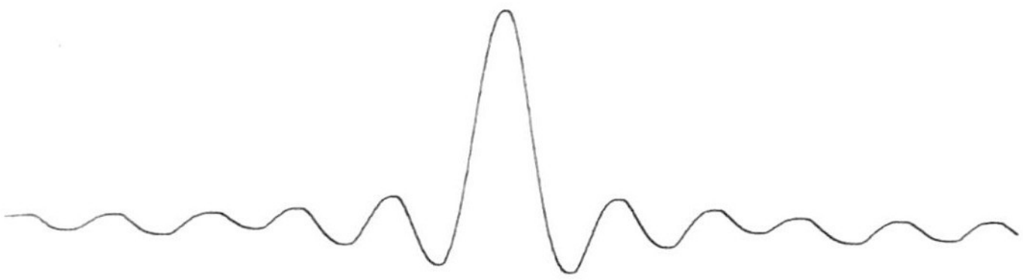
... and produces this output on the front of the machine.

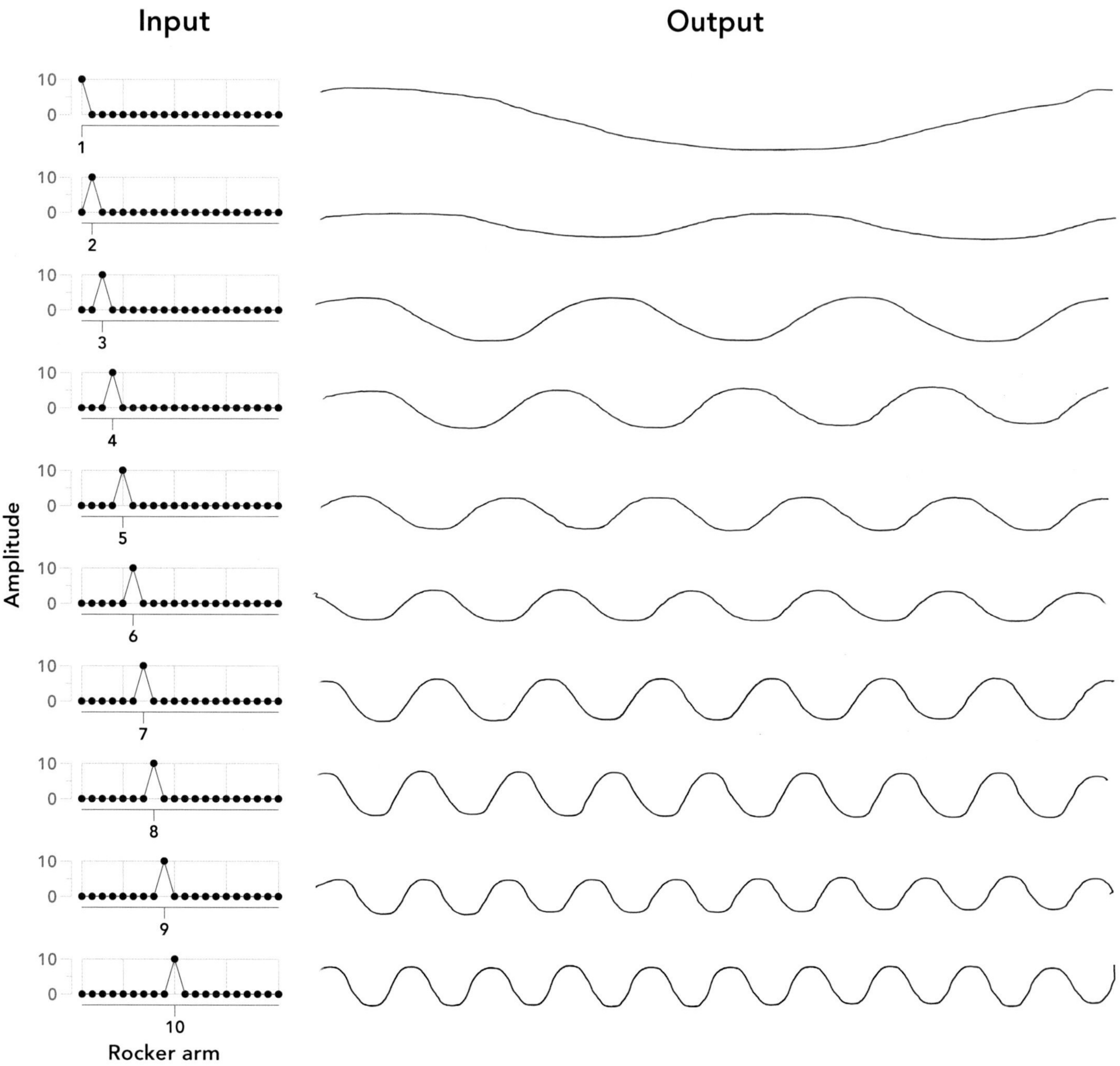
Input
Output
Amplitude
10
0
1
2
3
4
5
6
7
8
9
10
Rocker arm

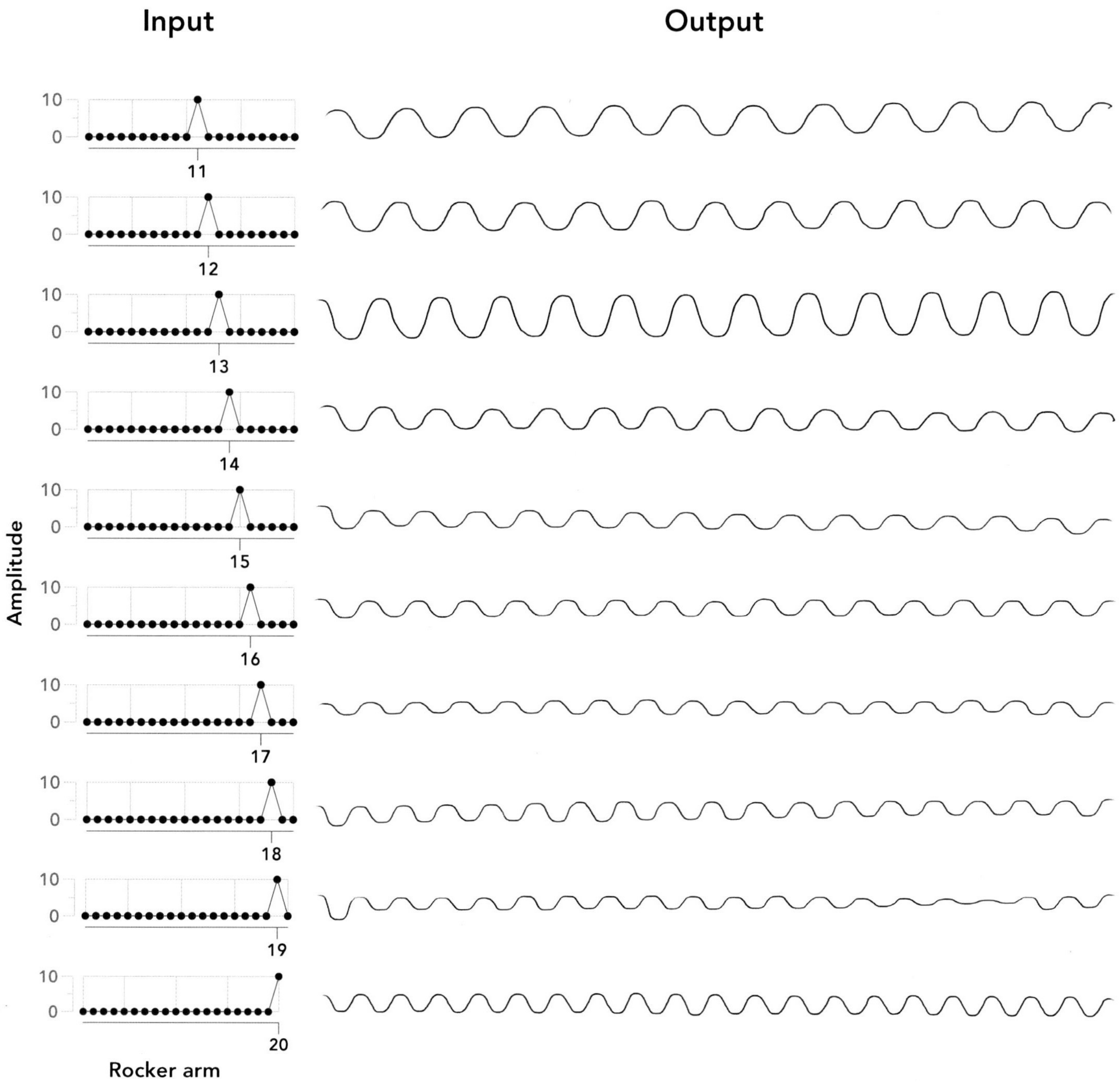
Input
Output
Amplitude
10
0
11
12
13
14
15
16
17
18
19
20
Rocker arm

Input

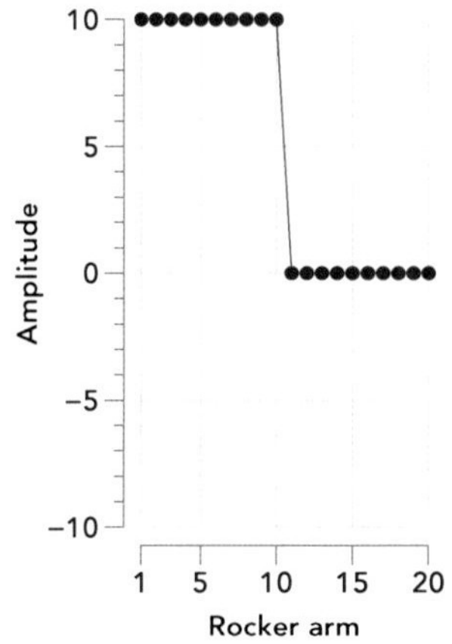
10
5
0
−5
−10
Amplitude
1 5 10 15 20
Rocker arm

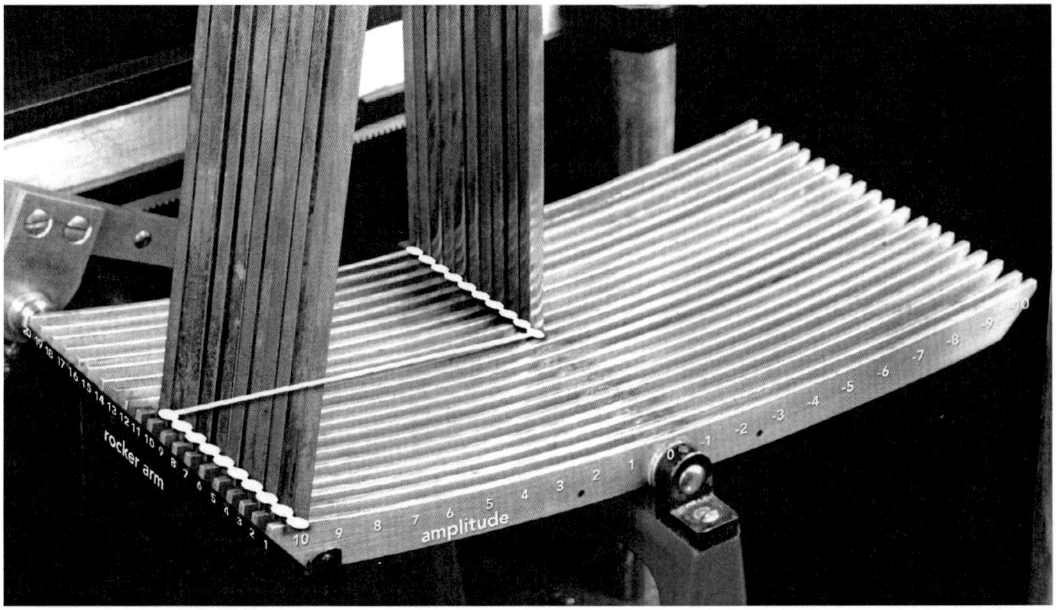
rocker arm
amplitude

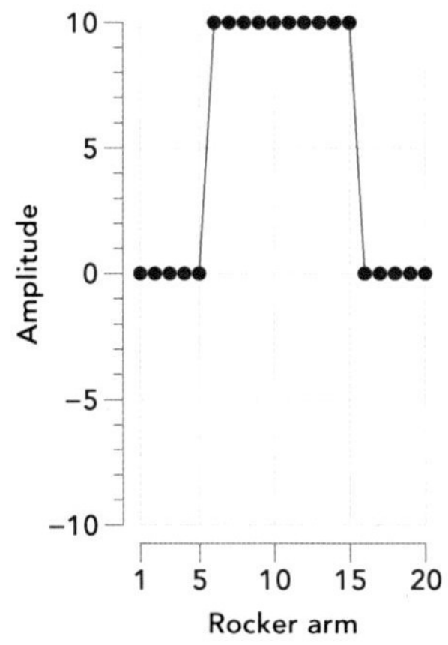
10
5
0
−5
−10
Amplitude
1 5 10 15 20
Rocker arm

Output

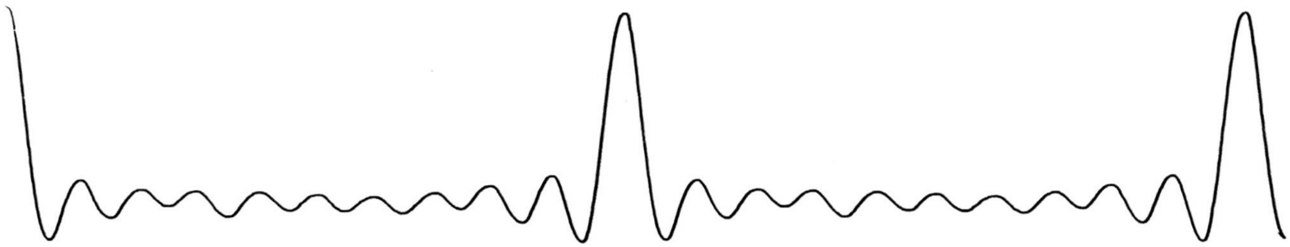

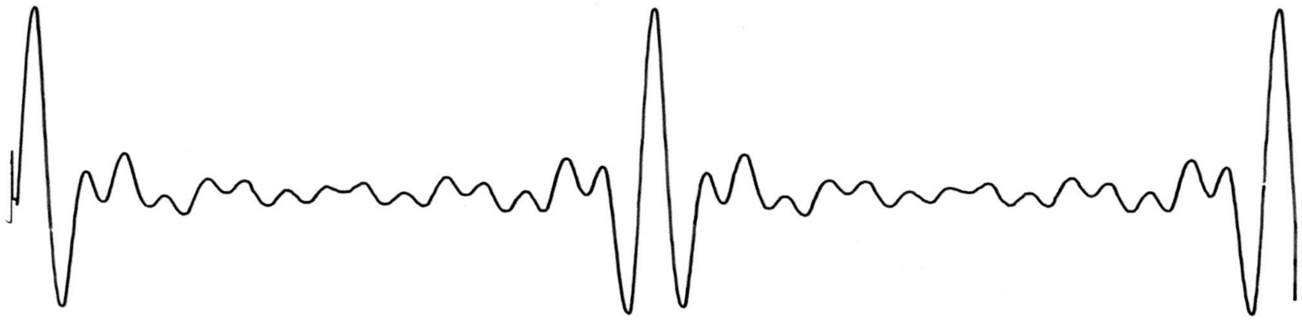

Input

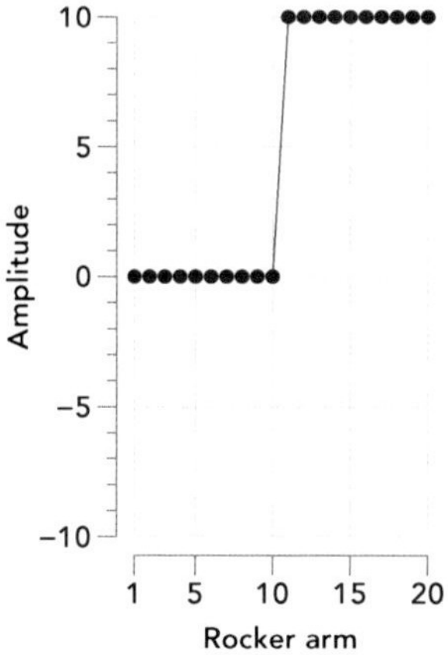
10
5
0
−5
−10
Amplitude
1
5
10
15
20
Rocker arm

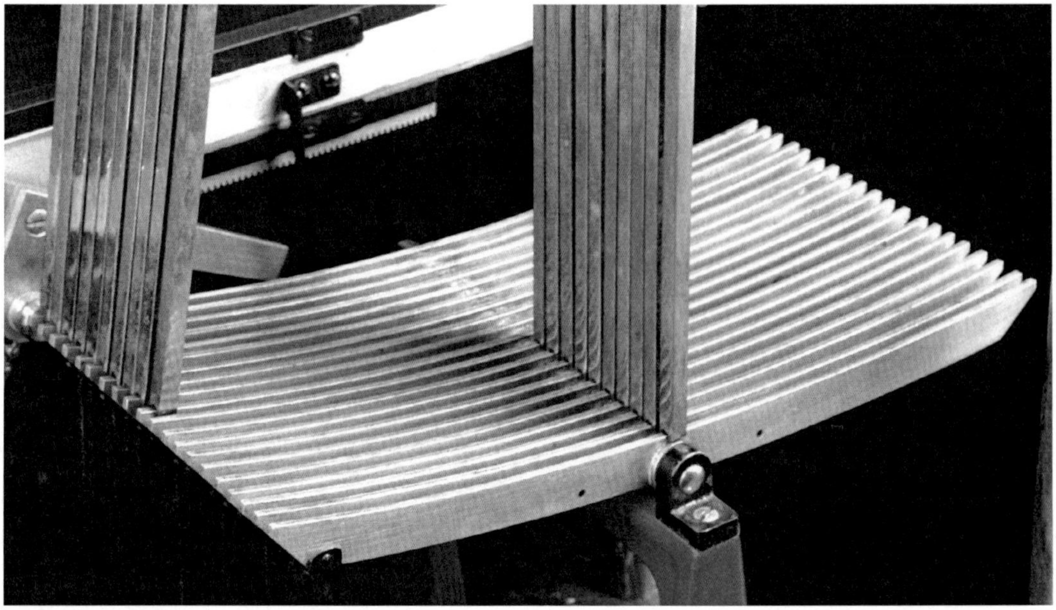

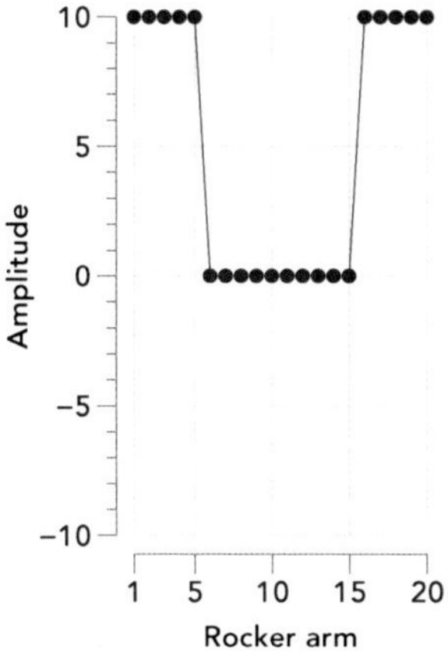
10
5
0
−5
−10
Amplitude
1
5
10
15
20
Rocker arm

Output

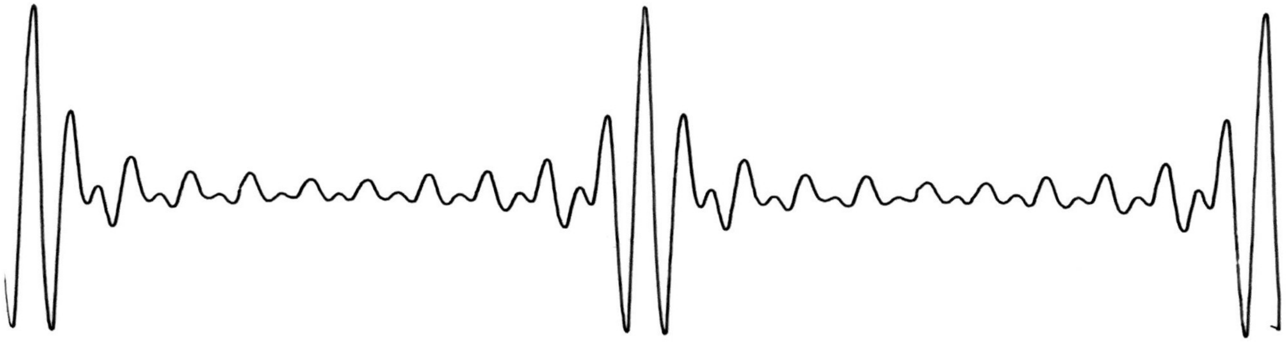

Input

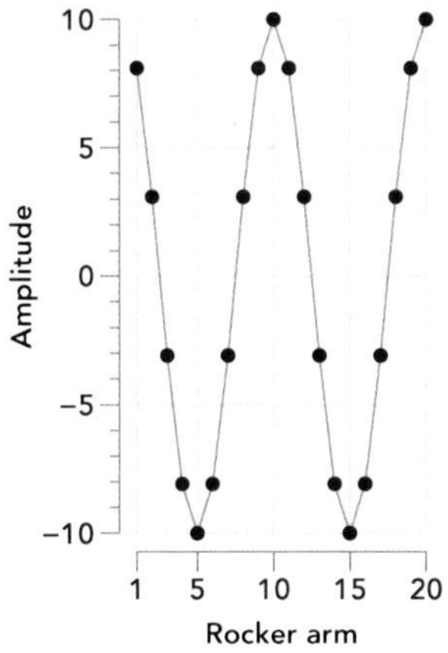
Amplitude
10
5
0
−5
−10
1
5
10
15
20
Rocker arm

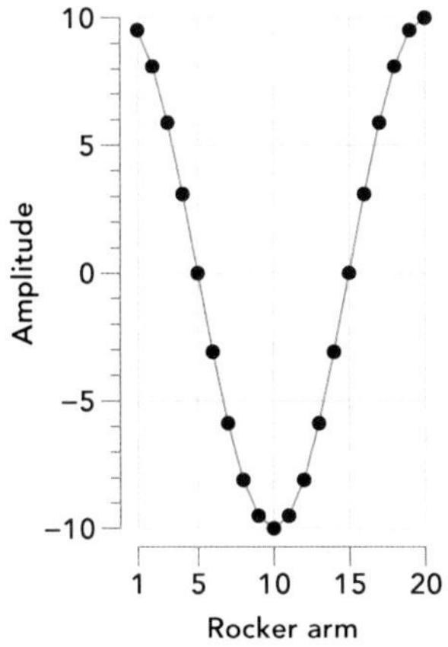
Amplitude
10
5
0
−5
−10
1
5
10
15
20
Rocker arm

Output

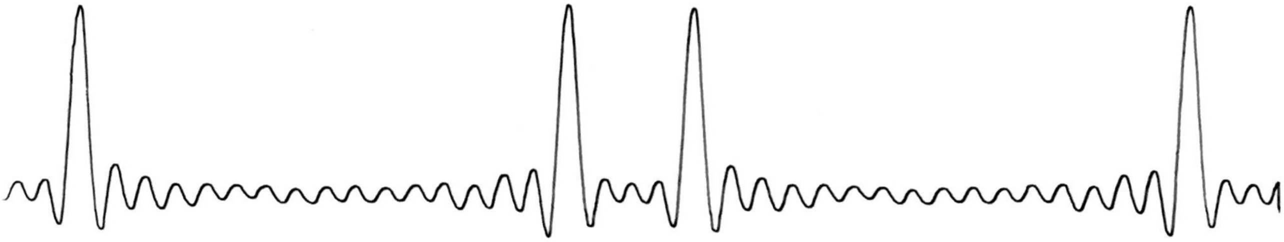

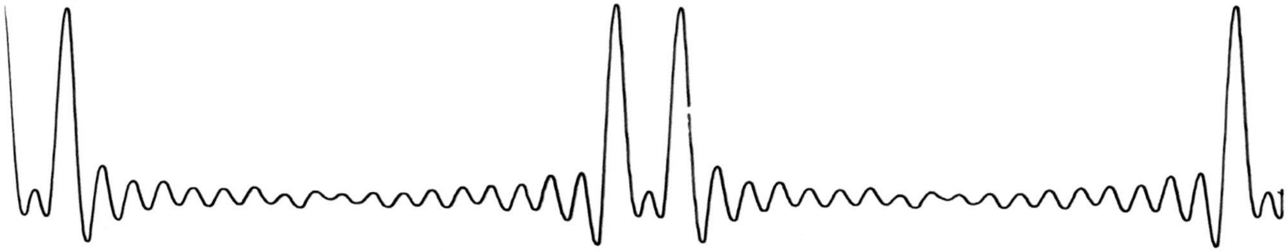

Input

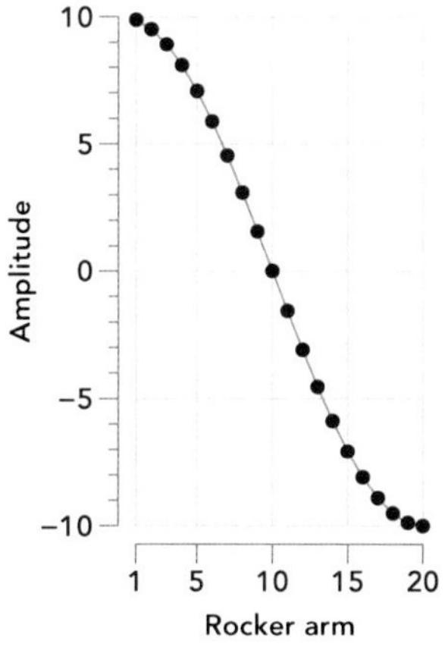
10
5
0
−5
−10
Amplitude
1
5
10
15
20
Rocker arm

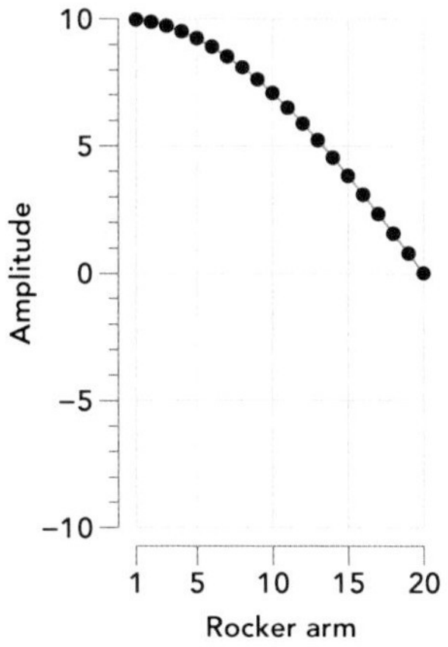
10
5
0
−5
−10
Amplitude
1
5
10
15
20
Rocker arm

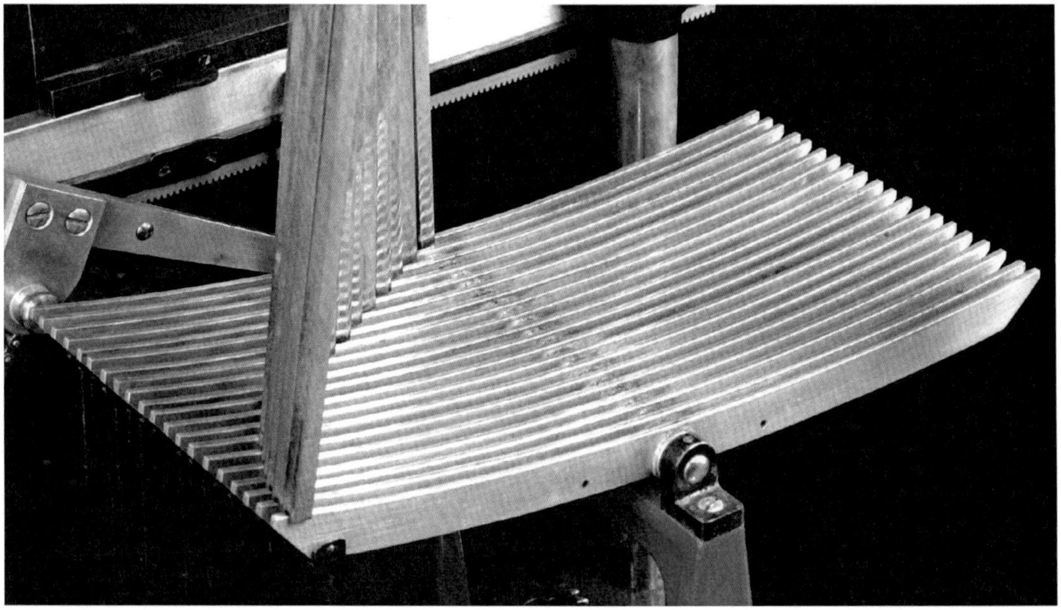

Output

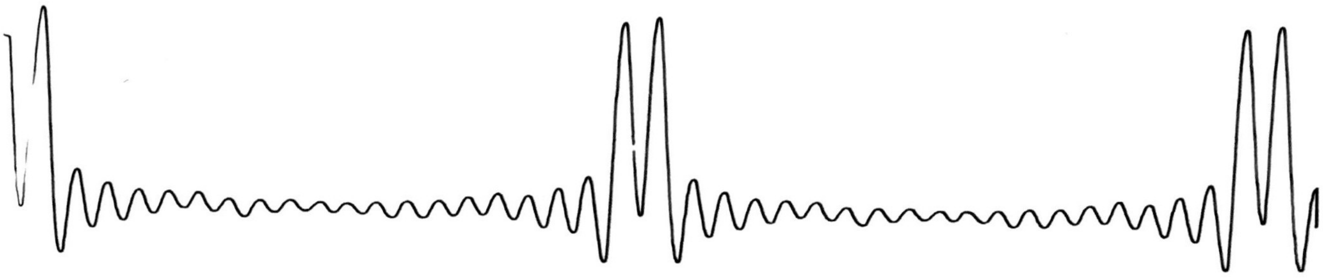

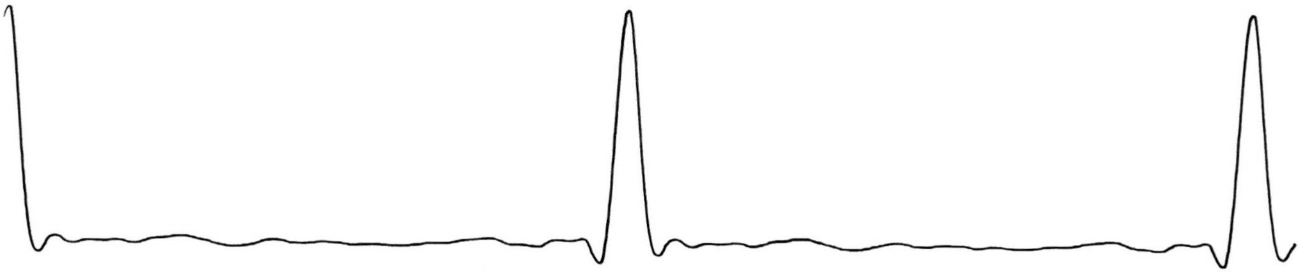

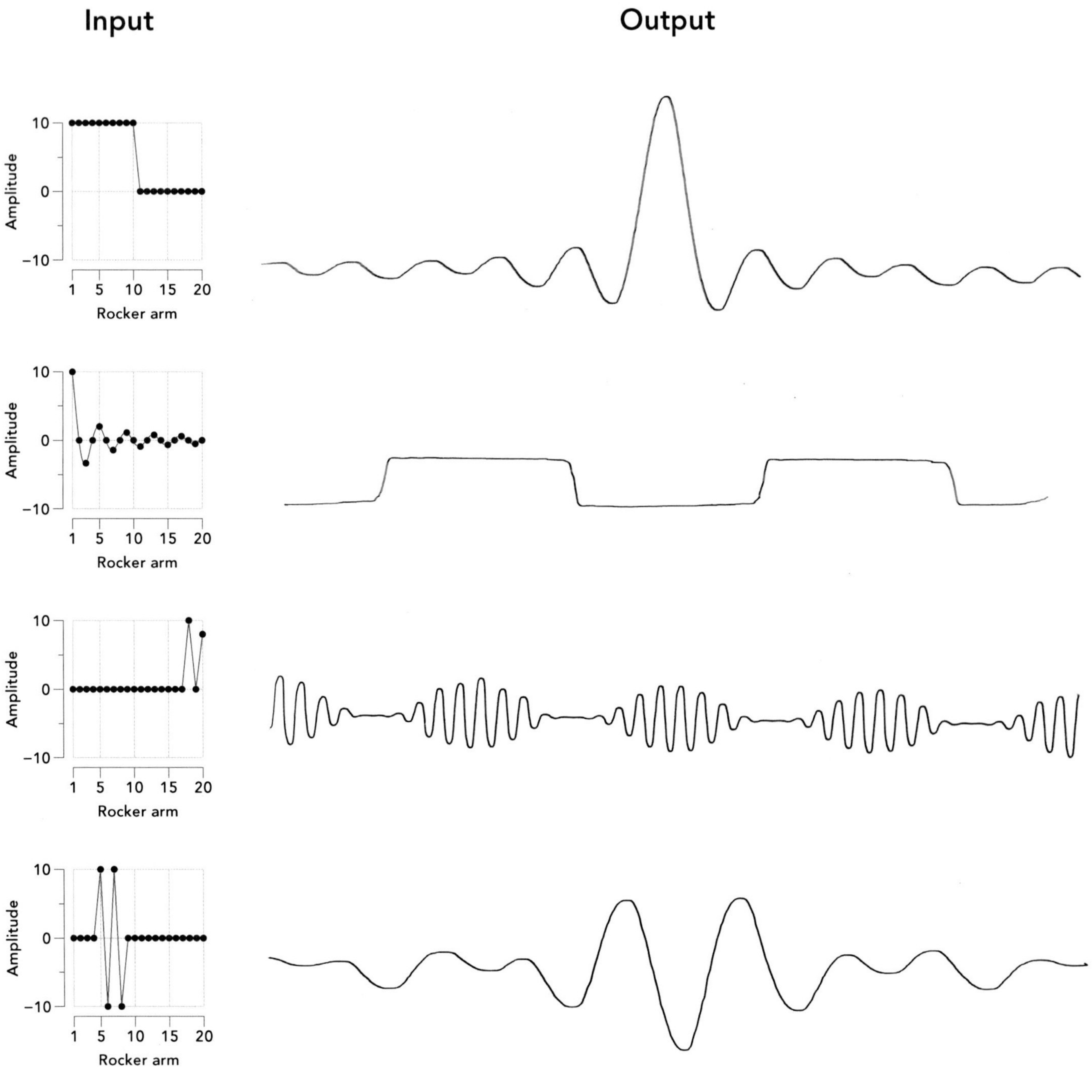
Input
Output
10
0
−10
Amplitude
1 5 10 15 20
Rocker arm
10
0
−10
Amplitude
1 5 10 15 20
Rocker arm
10
0
−10
Amplitude
1 5 10 15 20
Rocker arm
10
0
−10
Amplitude
1 5 10 15 20
Rocker arm

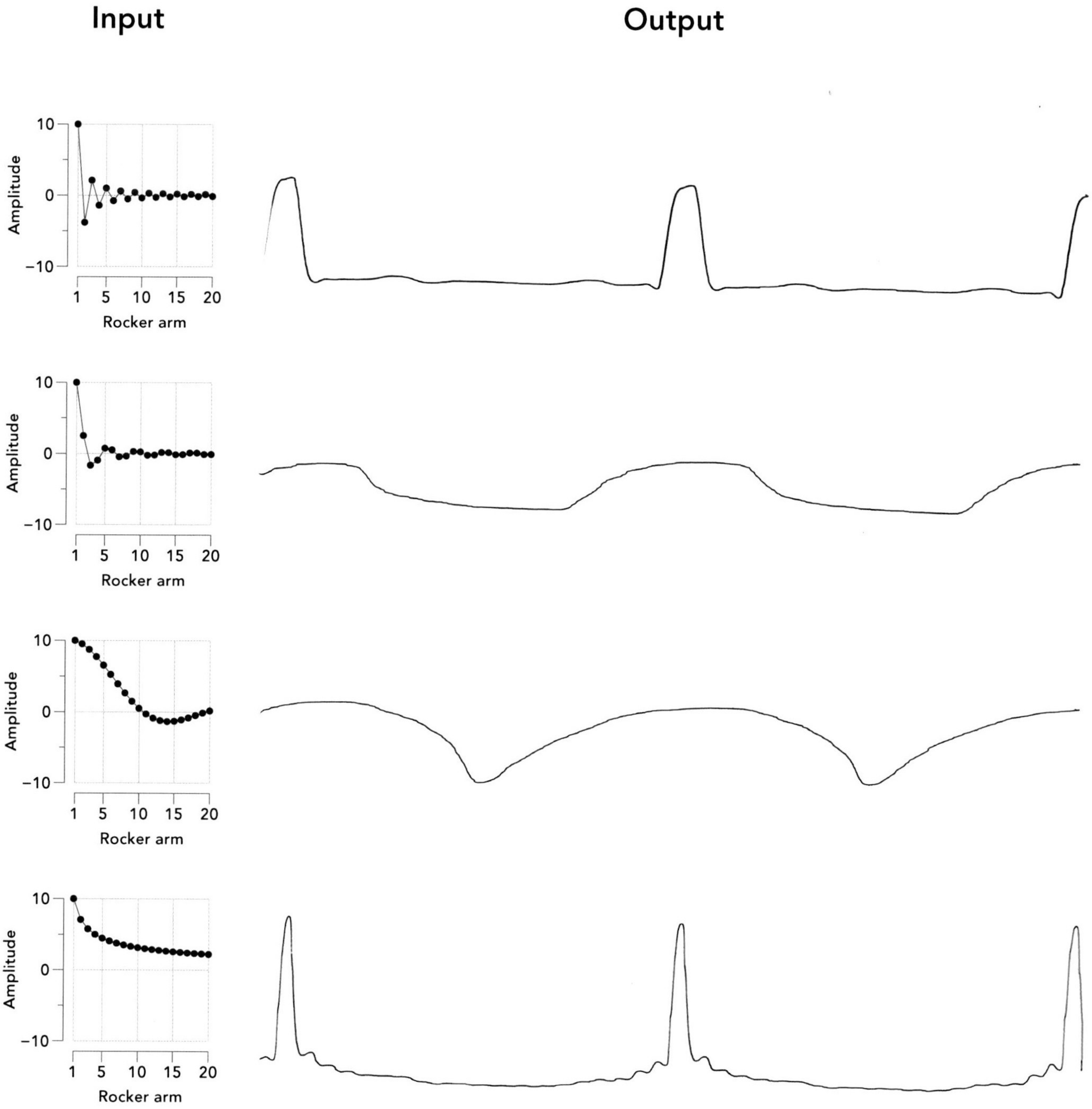
Input
Output
10
0
−10
Amplitude
1 5 10 15 20
Rocker arm
10
0
−10
Amplitude
1 5 10 15 20
Rocker arm
10
0
−10
Amplitude
1 5 10 15 20
Rocker arm
10
0
−10
Amplitude
1 5 10 15 20
Rocker arm

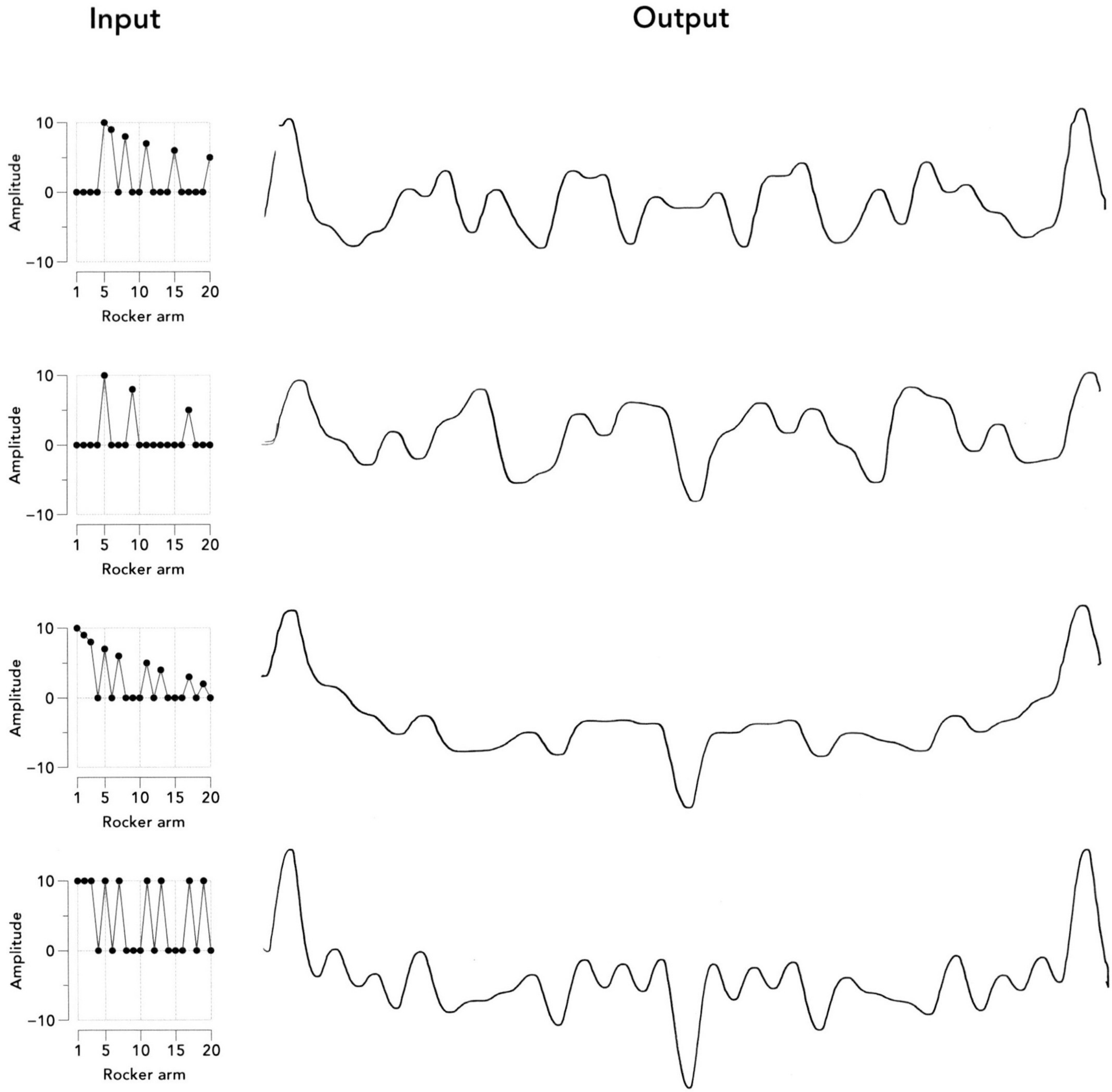
Input
Output
10
0
−10
Amplitude
1 5 10 15 20
Rocker arm

Input
machine set for sines

Output
machine set for sines

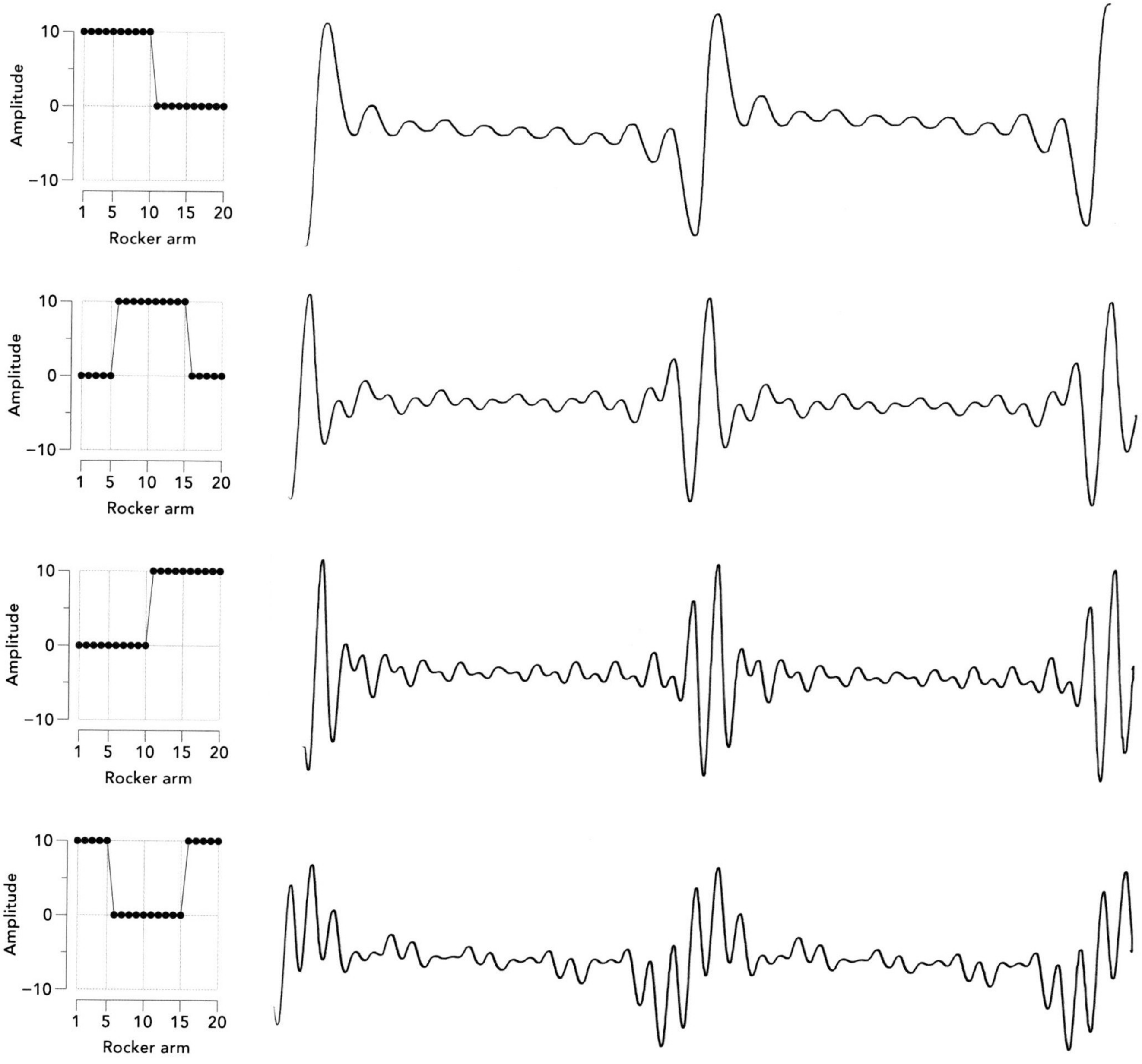

Michelson's 1898 paper

A. A. Michelson and S. W. Stratton
"A New Harmonic Analyzer"
American Journal of Science 25 (1898): 1-13

In this paper Michelson and his coauthor Samuel Stratton describe an 80-element analyzer—a machine with four times as many elements as the analyzer shown in this book. They outline the essential mechanical elements of the analyzer, show pages of sample output, and take a brief look at the mathematical approximations and simplifications underlying the machine's operation. A close look at the paper will delight the reader. For example, the function shown on the left and right sides of figure 13 is the profile of a human face. And, at the end of the paper, Michelson and Stratton propose two intriguing ideas. First, they propose building a machine with 1000 elements. Second, they note that the sinusoidal motions created by the gears could be replaced by other functions.

THE

AMERICAN JOURNAL OF SCIENCE

[FOURTH SERIES.]

ART. I.—*A new Harmonic Analyzer*; by A. A. MICHELSON and S. W. STRATTON. (With Plate I.)

EVERY one who has had occasion to calculate or to construct graphically the resultant of a large number of simple harmonic motions has felt the need of some simple and fairly accurate machine which would save the considerable time and labor involved in such computations.

The principal difficulty in the realization of such a machine lies in the accumulation of errors involved in the process of addition. The only practical instrument which has yet been devised for effecting this addition is that of Lord Kelvin. In this instrument a flexible cord passes over a number of fixed and movable pulleys. If one end of the cord is fixed, the motion of the other end is equal to twice the sum of the motions of the movable pulleys. The range of the machine is however limited to a small number of elements on account of the stretch of the cord and its imperfect flexibility, so that with a considerable increase in the number of elements the accumulated errors due to these causes would soon neutralize the advantages of the increased number of terms in the series.

It occurred to one of us some years ago that the quantity to be operated upon might be varied almost indefinitely, and that most of the imperfections in existing machines might be practically eliminated. Among the methods which appeared most promising were addition of fluid pressures, elastic and other forces, and electric currents. Of these the simplest in practice is doubtless the addition of the forces of spiral springs.

The principle upon which the use of springs depends may be demonstrated as follows:

Let a (Fig. 1)=lever arm of small springs, s. (but one of which is shown in the fig.)
b=lever arm of large counter-spring, S.
l_0=natural length of small springs.
L_0=natural length of large springs.
$l+x$=stretched length of small springs.
$L+y$=stretched length of large springs.
e=constant of small springs.
E=constant of large springs.
n=number of small springs.
p=force due to one of the small springs.
P=force due to the large spring.

then

$$p=\frac{e}{l_0}(l+x-\frac{a}{b}y)$$

$$P=\frac{E}{L_0}(L+y)$$

$$a\Sigma p=bP.$$

whence

$$y=\frac{\Sigma x}{n\left(\frac{l}{L}+\frac{a}{b}\right)}$$

From this it follows that the resultant motion is proportional to the algebraic sum of the components, at least to the same order of accuracy as the increment of force of every spring is proportional to the increment of length.

To obtain the greatest amplitude for a given number of elements, the ratios $\frac{l}{L}$ and $\frac{a}{b}$ should be as small as possible, but of course a limit is soon reached, when other considerations enter.

About a year ago a machine was constructed on this principle with twenty elements and the results obtained* were so encouraging that it was decided to apply to the Bache Fund for assistance in building the present machine of eighty elements.

Fig. 1 shows the essential parts of a single element. s is one of eighty small springs attached side by side to the lever C, which for greater rigidity has the form of a hollow cylinder, pivoted on knife edges at its axis. S is the large counter-spring. The harmonic motion produced by the excentric A, is communicated to x by the rod R and lever B, the amplitude of the motion at x depending on the adjustable distance d.

* Paper read before the National Academy of Science, April, 1897.

1.

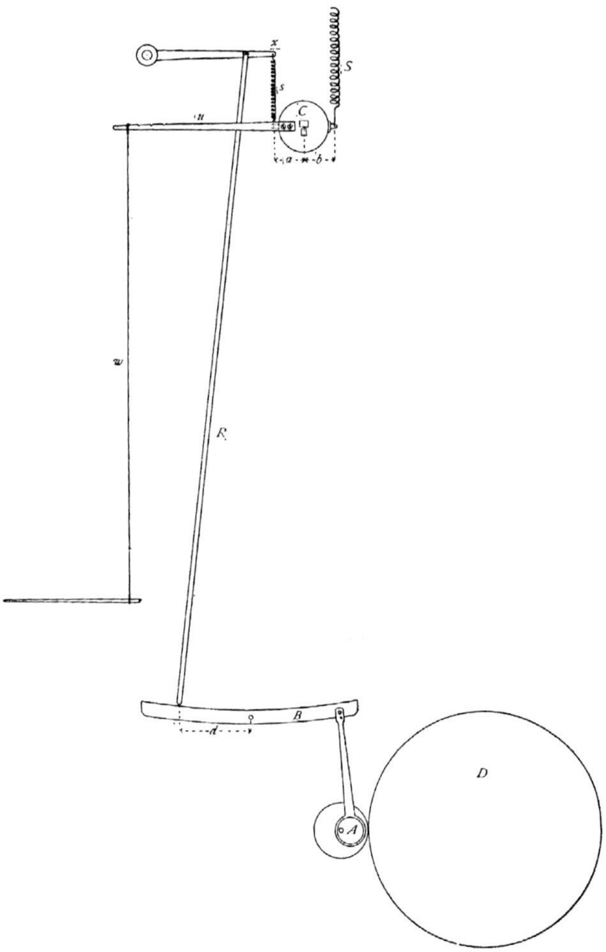

The resultant motion is recorded by a pen connected with u by a fine wire w. Under the pen a slide moves with a speed proportional to the angular motion of the cone D. (Plate I.)

To represent the succession of terms of a Fourier series the excentrics have periods increasing in regular succession from one to eighty. This is accomplished by gearing to each excentric a wheel, the number of whose teeth is in the proper ratio. These last are all fastened together on the same axis and form the cone D. (Plate I.)

2.

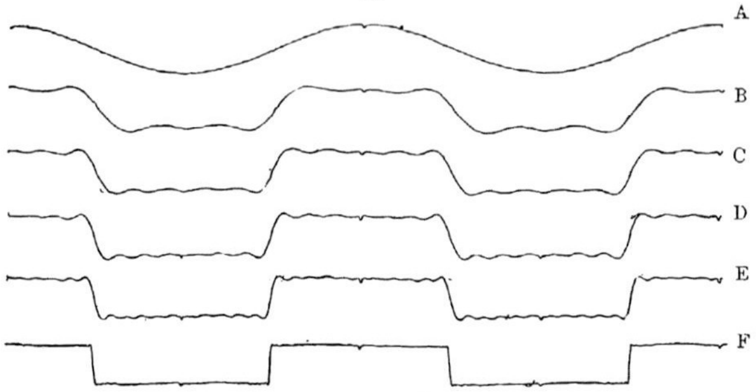

A, one term; B, five terms; C, nine terms; D, thirteen terms; E, twenty one terms; F, seventy-nine terms.

Turning the cone will produce at the points (x) motions corresponding to $\cos\theta$, $\cos 2\theta$, $\cos 3\theta$, etc., up to $\cos 80\theta$, and whose amplitudes depend on the distances d. The motion of the elements may also be changed from sine to cosine by disengaging the cone and turning all of the excentrics through 90° by means of a long pinion which can be thrown in gear with all of the excentric wheels at once.

The efficiency and accuracy of the machine is well illustrated in the summation of Fourier series shown in the accompanying figures.

Figure 2 shows the dependence of the accuracy of a particular function on the number of terms of the series. Figures 3, 4, 5, 6 and 7 are illustrations of a number of standard forms, and 8, 9 and 10 illustrate the use of the machine in constructing curves representing functions which scarcely admit of other analytical expression.

The machine is capable not only of summing up any given trigonometrical series but can also perform the inverse process

3.

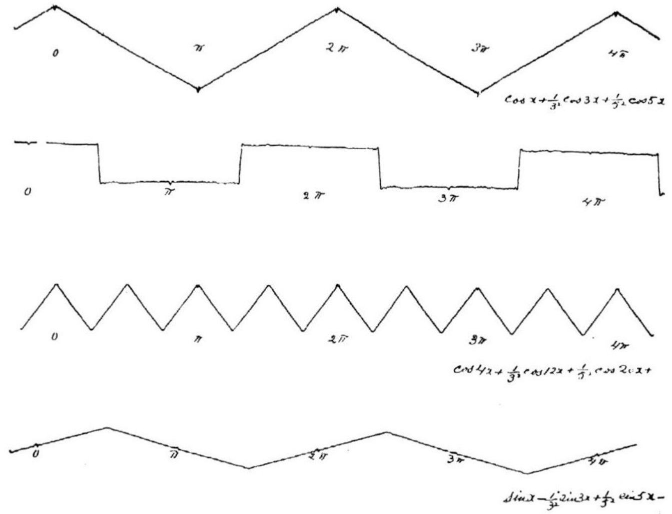

4.

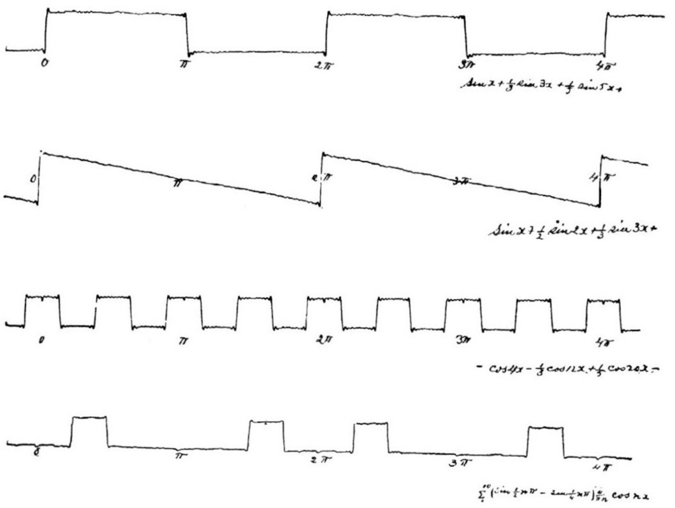

5.

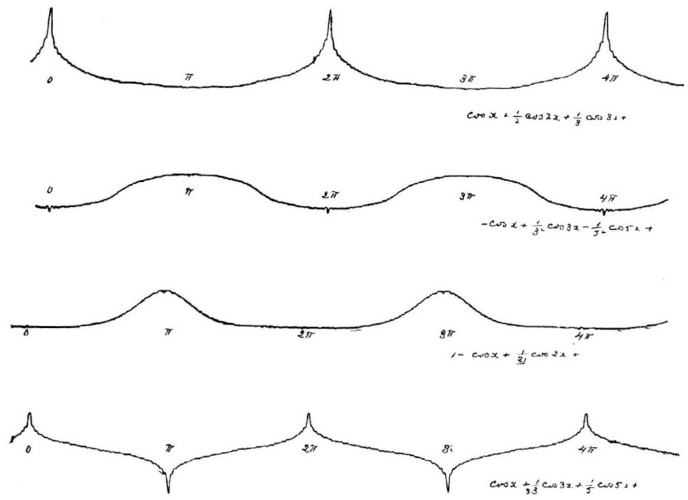

6.

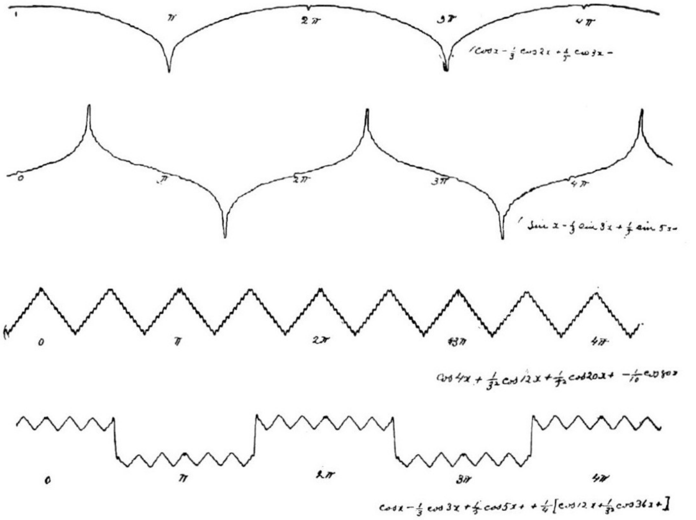

7.

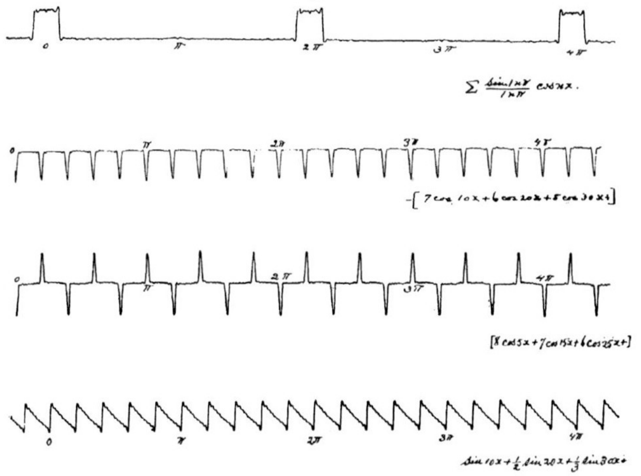
[8 cos 5x + 7 cos 15x + 6 cos 25x +]
sin 10x + ½ sin 20x + ⅓ sin 30x +

8.

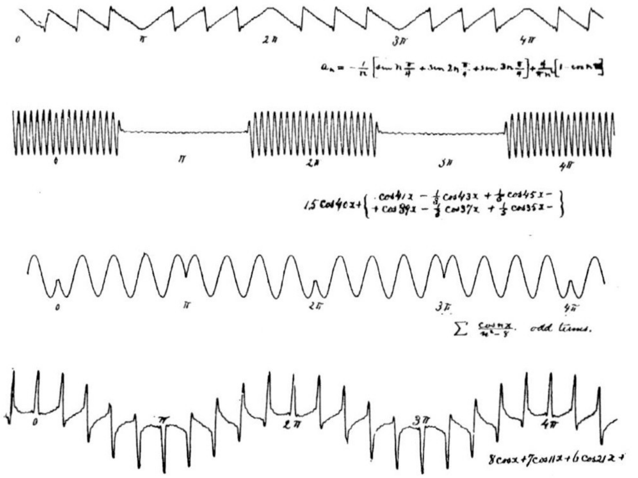

9.

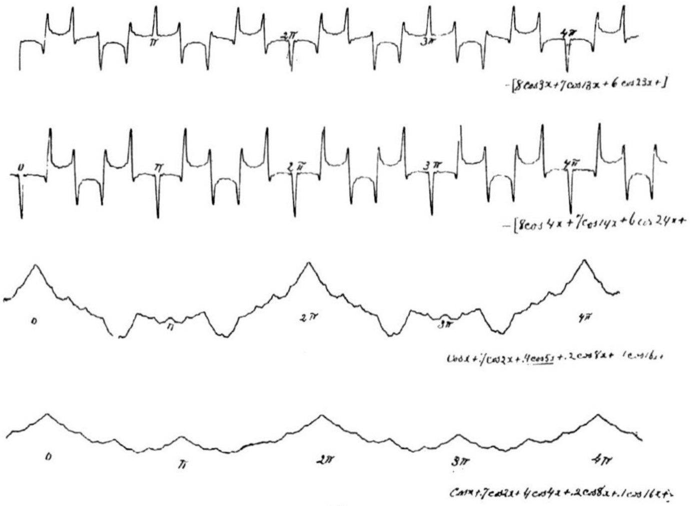
−[8 cos 3x + 7 cos 13x + 6 cos 23x +]
−[8 cos 4x + 7 cos 14x + 6 cos 24x +

10.

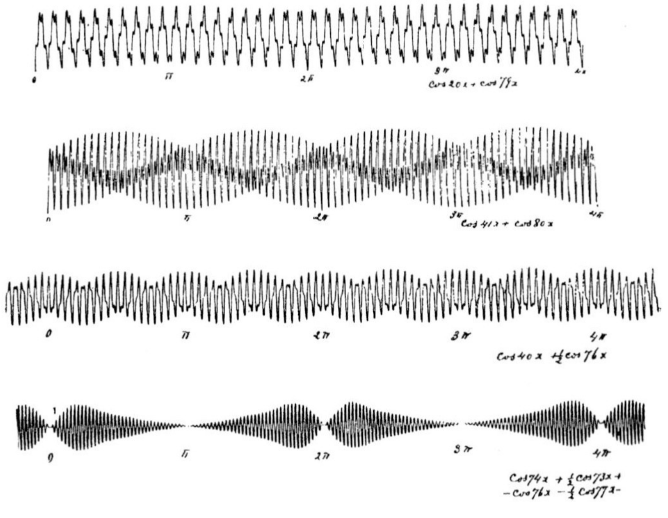

of finding for any given function the coefficients of the corresponding Fourier series.

Thus if $f(x)=a_0+a_1 \cos x+a_2 \cos 2x+$

we have $a_k=\frac{2}{\pi}\int_0^{\pi} f(x) \cos kx\, dx$

11.

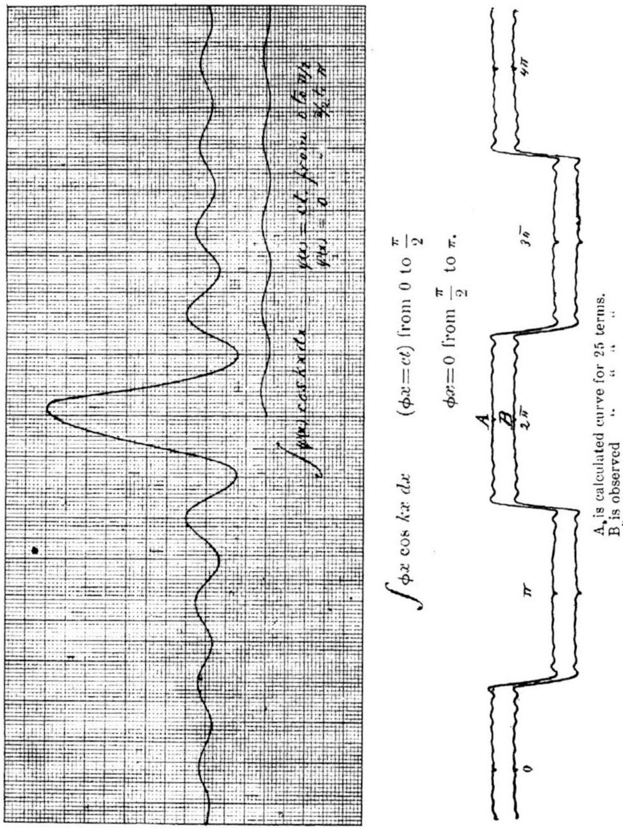

On the other hand, if n is the number of an element of the machine and a the distance between any two elements, and the

amplitude d (fig. 1) is proportional to $f(na)$, the machine gives

$$\sum_0^m f(na) \cos n\theta = \sum_0^{\pi} f(x) \cos \frac{m}{\pi}\theta x$$

which is proportional to a_k if $k=\frac{m}{\pi}\theta$. Hence to obtain the integral, the lower ends of the vertical rods R (Plate I) are moved along the levers B to distances proportional to the ordinates of the curve $y=f(na)$.

The curve thus obtained for a_k is a *continuous* function of k which approximates to the value of the integral as the number of elements increases. To obtain the values corresponding to the coefficients of the Fourier series, the angle $\theta=\pi$, or the corresponding distance on the curve, is divided into m equal parts. The required coefficients are then proportional to the ordinates erected at these divisions.

Figure 11 gives the approximate value of $\int \varphi(x) \cos kx\, dx$ when $\varphi(x)=$constant from 0 to a, and is zero for all other values. The exact integral is $\frac{\sin ka}{k}$. The accuracy of the approximation is shown by the following table, which gives the observed and the calculated values of the first twenty coefficients for $a=4{\cdot}0$.

$$\int_0^a \cos kx\, dx$$

n.	obs.	calc.	△
0	100·0	100·0	0·0
1	65·0	64·0	1·0
2	0·0	0·0	0 0
3	—20·0	—21·0	1·0
4	0·0	0·0	0·0
5	12·5	13·0	—0·5
6	—1·5	0·0	—1·5
7	—9·0	—9·0	0 0
8	0·0	0·0	0·0
9	6 0	7·0	—1·0
10	0·0	0·0	—2·0
11	—6·0	—6·0	0·0
12	0·0	0·0	—0.0
13	4·0	5·0	—1·0
14	—2·0	0.0	—2·0
15	—4·0	4·5	0·5
16	0·5	0·0	0·5
17	8·5	4·0	—0·5
18	—1·0	0·0	—1·0
19	—3 5	—3·0	0·5
20	0·0	0·0	0·0

The average error is only 0·65 of one per cent. of the value of the greatest term.

The accuracy of the result is also shown in curves A and B (fig. 11). The former gives the summation of the calculated terms and the latter of the observed.

Another illustration is given in figure 12 in which

$\varphi(x) = e^{-a^2x^2}$

For $a = \cdot 1$ the following are the values of the coefficients of the first twelve terms of the equivalent Fourier series.

$$\int_0^\infty e^{-a^2x^2} \cos kx\, dx$$

n.	obs.	calc.	△
0	100·0	100·0	0·0
1	95·0	96·0	—1·0
2	85·0	86·0	—1·0
3	70·0	70·0	0·0
4	53·0	54·0	—1.0
5	38·0	38·0	0·0
6	25·0	25·0	0·0
7	16·0	15·0	1·0
8	8·8	8·0	—0·8
9	5·0	4·5	0·5
10	3·6	2·0	1·6
11	2·4	1·0	1·4
12	1·6	0·5	1·1

Here the average error is only 0·7 per cent of the value of the greatest term.

The complete cycle of operations of finding the coefficients of the complete Fourier series (sines and cosines) and their recombination, reproducing the original function, is illustrated in figure 13. *A* is the original curve. *B* and *C* are the values of $\int\varphi(x) \sin kx\, dx$ and $\int\varphi(x) \cos kx\, dx$ respectively. Their in-

12.

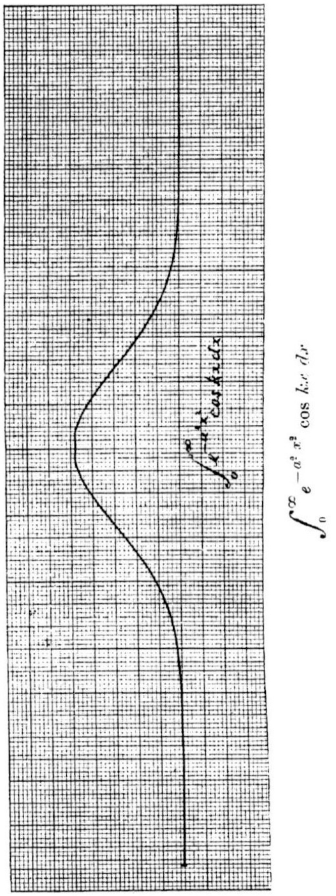

13.

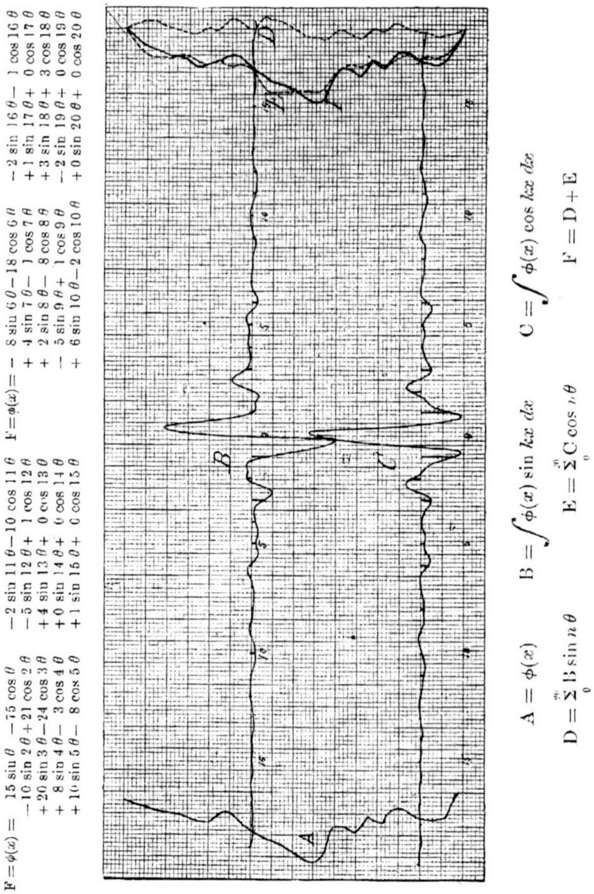

tersections with the ordinates midway between the heavier ordinates, give the coefficients of the sine and cosine series respectively. The sums of the first twenty terms are represented by the curves D and E and finally the sum of these two curves produces the curve F, which agrees sufficiently well with the original to be easily recognizable.

It appears, therefore, that the machine is capable of effecting the integration $\int\varphi(x)\cos kx\,dx$ with an accuracy comparable with that of other integrating machines; and while it is scarcely hoped that it will be used for this purpose where great accuracy is required, it certainly saves an enormous amount of labor in cases where an error of one or two per cent is unimportant.

The experience gained in the construction of the present machine shows that it would be quite feasible to increase the number of elements to several hundred or even to a thousand with a proportional increase in the accuracy of the integrations.

Finally it is well to note that the principle of summation here employed is so general that it may be used for series of any function by giving to the points (p) the motions corresponding to the required functions, instead of the simple harmonic motion furnished by the excentrics. A simple method of effecting this change would be to cut metal templates of the required forms, mounting them on a common axis. In fact the harmonic motion of the original machine was thus produced.

Ryerson Physical Laboratory, University of Chicago.

A New Harmonic Analyzer.

Math Overview: Synthesis

THIS HARMONIC ANALYZER implements a simplified version of a mathematical technique pioneered in the early 1800s by Joseph Fourier. Many periodic functions can be represented by a series of cosines and sines:

$$f(x) = \frac{a_0}{2} + \sum_{n=1}^{\infty} \left[a_n \cos\left(\frac{2\pi\, n\, x}{T}\right) + b_n \sin\left(\frac{2\pi\, n\, x}{T}\right)\right]$$

where a_0, a_n and b_n are constants, and T is the period.

The analyzer can be set up to synthesize either an *even* periodic or an *odd* periodic function. A function is odd if when rotated 180° about the origin, the rotated function is identical to the unrotated function. In mathematical terms this occurs when:

$$f(-x) = -f(x)$$

A function is even if when mirrored about the vertical axis, the mirrored function is identical to the unmirrored function. In mathematical terms this occurs when:

$$f(-x) = f(x)$$

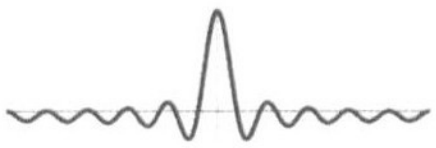

An odd periodic function can be approximated using only sines:

$$f(x) = \sum_{n=1}^{\infty} b_n \sin\left(\frac{2\pi\, n\, x}{T}\right)$$

while an even periodic function can be approximated using only cosines:

$$f(x) = \frac{a_0}{2} + \sum_{n=1}^{\infty} a_n \cos\left(\frac{2\pi\, n\, x}{T}\right)$$

When performing synthesis the Michelson analyzer uses several simplifications and approximations. To explain, we'll use only the cosine series, although everything can be recast easily in terms of the sine series.

The leading term of the series ($a_0/2$) is set using a knob that slides the writing mechanism up or down relative to the platen separately from the the sum of cosines (pg. 48); this action mimics the effect of the leading term, which simply slides the function up or down the vertical axis. This allows us to simplify the formula for synthesis to:

$$f(x) = \sum_{n=1}^{\infty} a_n \cos\left(\frac{2\pi\, n\, x}{T}\right)$$

The next simplification involves rescaling the horizontal axis. This axis, on which x is measured, does not have a fixed unit. It can be changed by the translational gearing that drives the platen, so we can assume that the period T is 2π. The formula now becomes:

$$f(x) = \sum_{n=1}^{\infty} a_n \cos(nx)$$

And, finally, an approximation is introduced: the machine can sum only twenty cosines becauses its gear train has only twenty gears. This restricts the sum to run from 1 to 20. Using these simplifications and approximations, the function synthesized by the analyzer becomes

$$f(x) = \sum_{n=1}^{20} a_n \cos(nx)$$

Using sines and cosines to approximate a function touches on many fundamental issues of mathematics and so its history is rich and fascinating. An excellent and accessible introduction to Fourier analysis and its history can be found in P.J. Davis, R. Hersh, and E.A. Marchisotto, *The Mathematical Experience* (New York: Springer, 2012).

Math Overview: Analysis

For a periodic function $f(x)$ with period T, the goal of analysis is to find the coefficients a_n and b_n needed to represent this function as a sum of sines and cosines:

$$f(x) = \frac{a_0}{2} + \sum_{n=1}^{\infty} \left[a_n \cos\left(\frac{2\pi n x}{T}\right) + b_n \sin\left(\frac{2\pi n x}{T}\right)\right]$$

We make the same simplifications as we did previously for synthesis, including working only with cosines. So, for an even periodic function $f(x)$ our goal is to determine the coefficients a_n in this equation:

$$f(x) = \sum_{n=1}^{\infty} a_n \cos(nx)$$

To calculate these coefficients we use the formula:

$$a_n = \frac{2}{\pi}\int_0^{\pi} f(x)\cos(nx)\,dx \qquad \text{where } n = 1, 2, 3, \ldots$$

For each value of n, the integral determining a_n can be approximated by a finite sum. Because we are working with a 20-element analyzer, we divide the interval $[0, \pi]$ into 20 sub-intervals, each of width $\Delta = \pi/20$:

$$a_n \approx \frac{2}{\pi}\sum_{k=1}^{20} f(k\Delta)\cos(n(k\Delta))\Delta \qquad \text{where } n = 1, 2, 3, \ldots, 20$$

$$a_n \approx \frac{2\Delta}{\pi}\sum_{k=1}^{20} f_k \cos(n(k\Delta)) \qquad \text{where } n = 1, 2, 3, \ldots, 20$$

where f_k denotes $f(k\Delta)$, the sampled value of the function at the kth sub-interval. We can ignore the leading factor of $2\Delta/\pi$ because we are concerned only with relative values of a_n. On the machine, these values can be scaled by adjusting the magnifying lever (pg. 46). This results in:

$$a_n \approx \sum_{k=1}^{20} f_k \cos\left(kn\frac{\pi}{20}\right) \qquad \text{where } n = 1, 2, 3, \ldots, 20$$

Notice that this is of the same form as the equation we use to synthesize a function with the machine! That is, it is the sum of weighted sinusoids. As the crank turns the machine produces continuous output, but in order to determine a_n, we are interested only in integer values of n. These integer values of n appear every two turns of the crank.

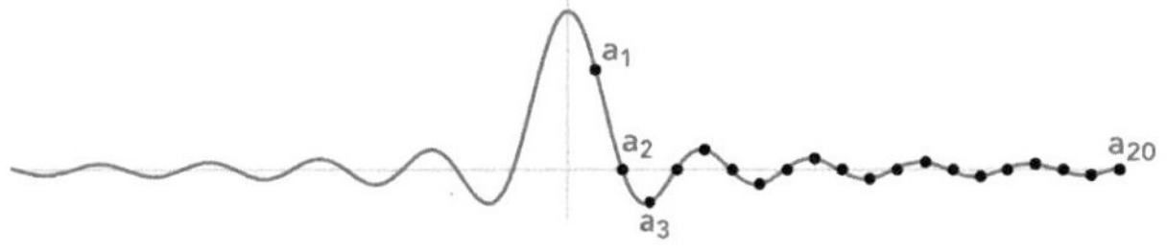

The analyzer's gears are sized such that a single full turn of the crank rotates the first gear of the cylindrical set through 1/80th of a full rotation, the second 2/80ths, the third 3/80ths, etc. This means that for two turns of the crank the first gear has rotated $\pi/20$, the second $2(\pi/20)$, and the third $3(\pi/20)$. Thus, two turns of the crank sets the cosine associated with the first gear to $\cos(\pi/20)$, the second to $\cos(2\pi/20)$, the third to $\cos(3\pi/20)$ and so on. This is the sequence of cosines used to approximate a_n when $n = 1$:

$$a_1 \approx f_1 \cos\left(\frac{\pi}{20}\right) + f_2 \cos\left(2\frac{\pi}{20}\right) + \ldots + f_{20}\cos\left(20\frac{\pi}{20}\right)$$

The other coefficients are approximated in the same way.

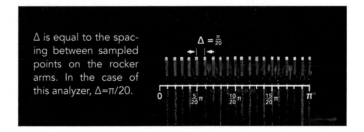

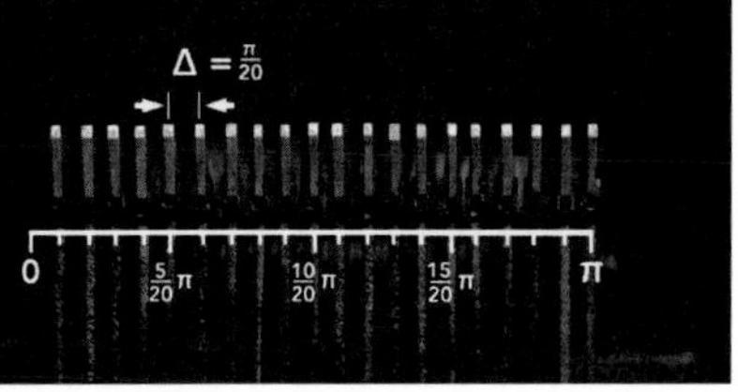

Δ is equal to the spacing between sampled points on the rocker arms. In the case of this analyzer, Δ=π/20.

Eight Views of the Machine

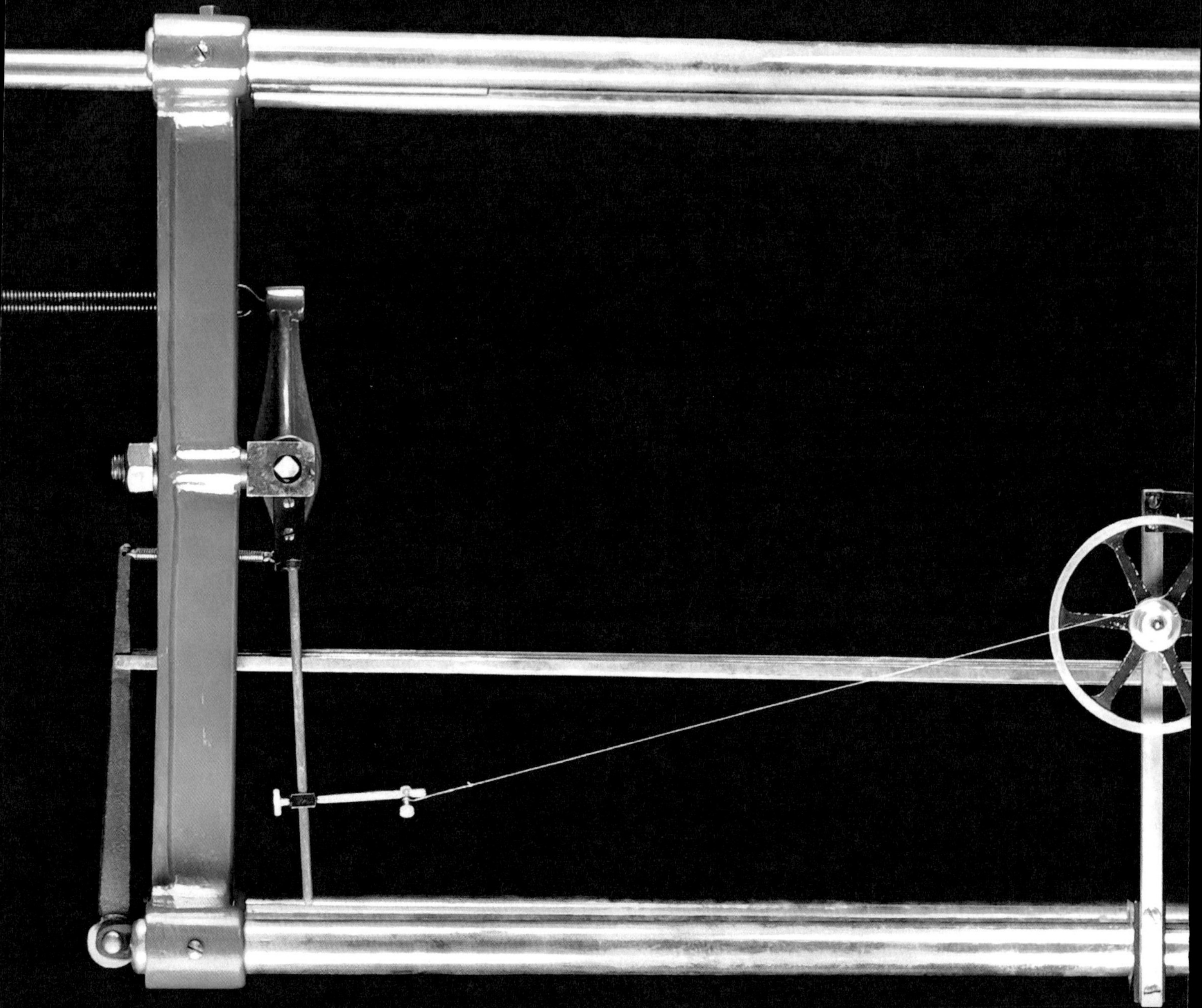

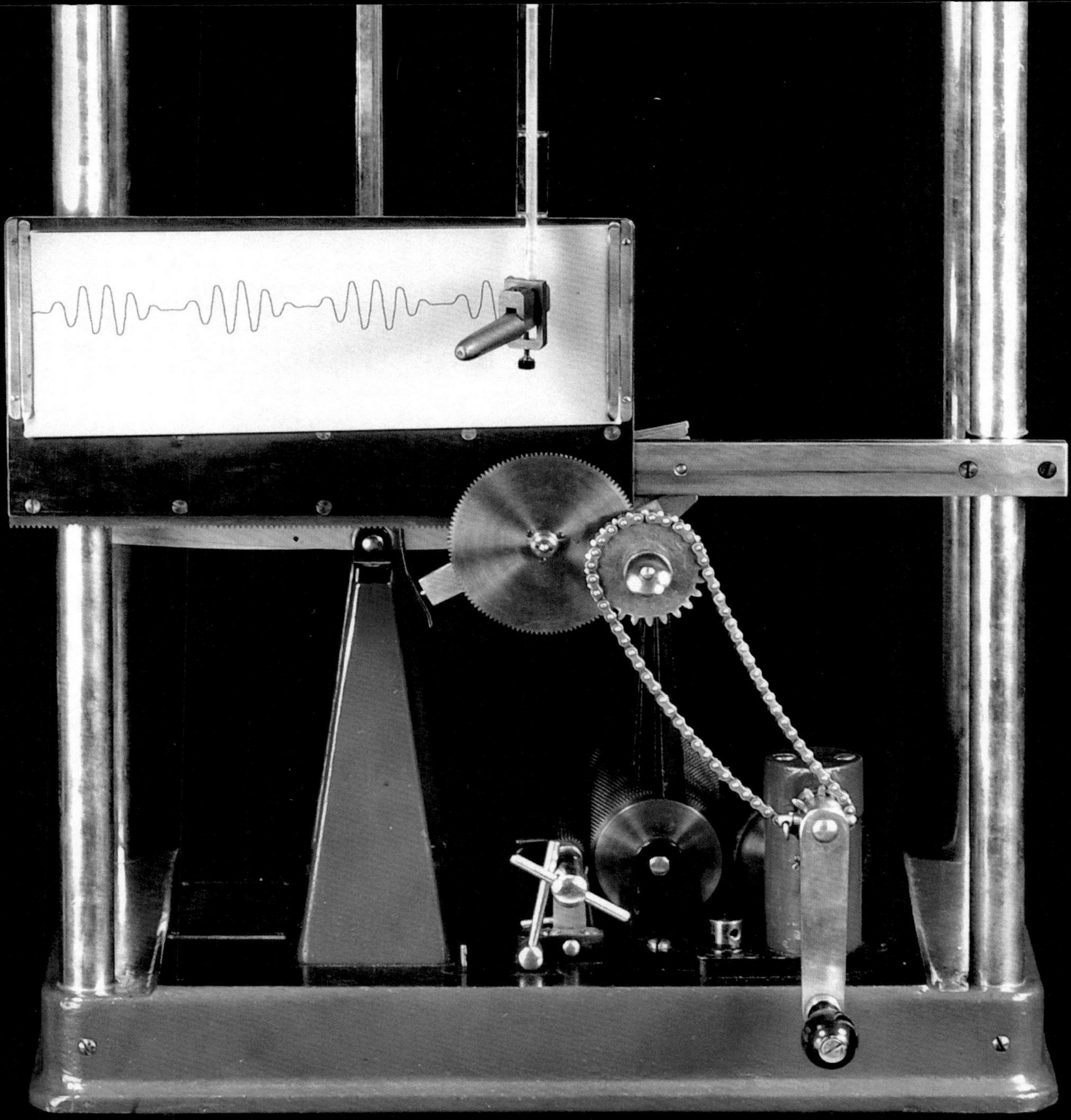

angle 0°
view 1/8

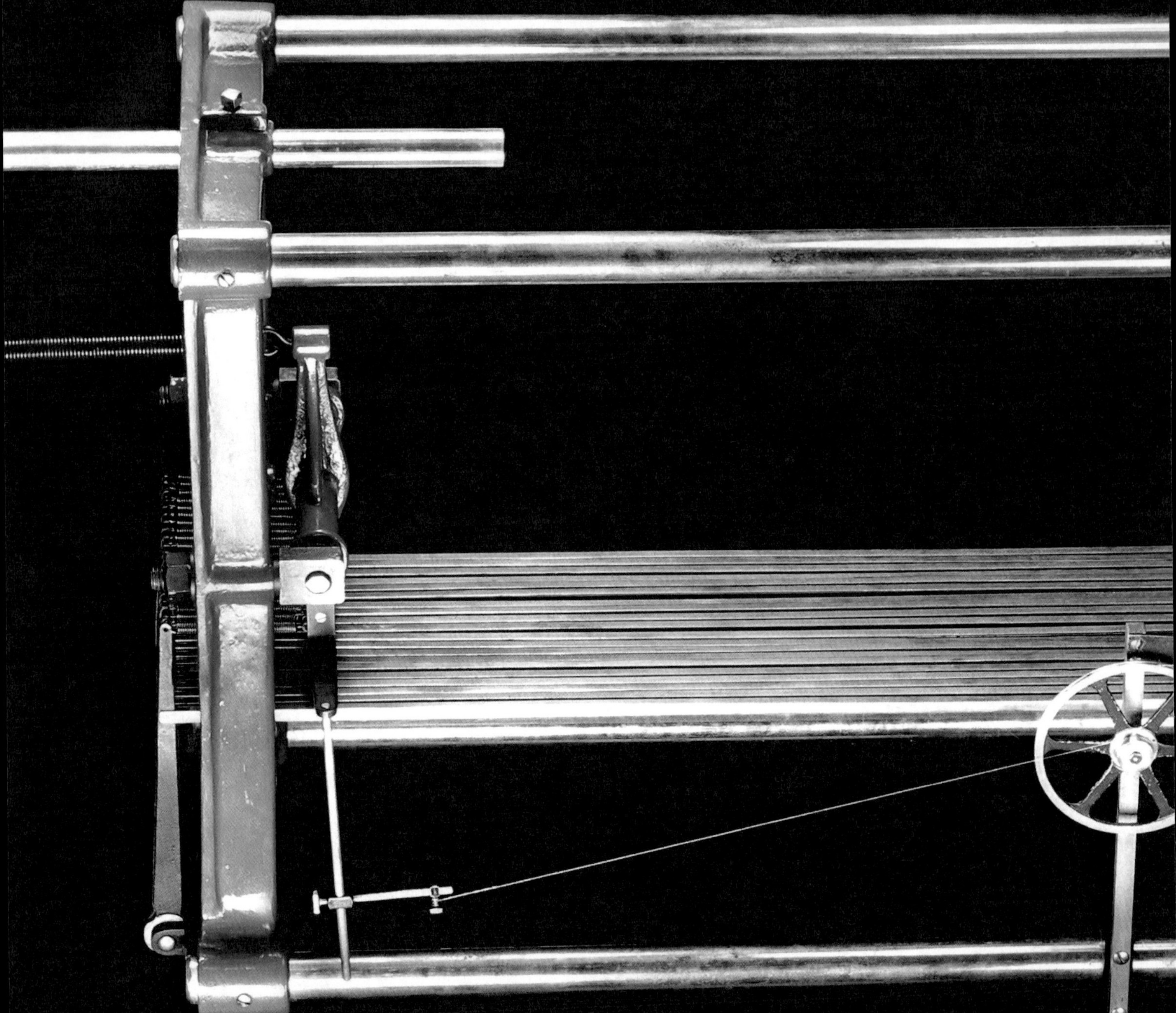

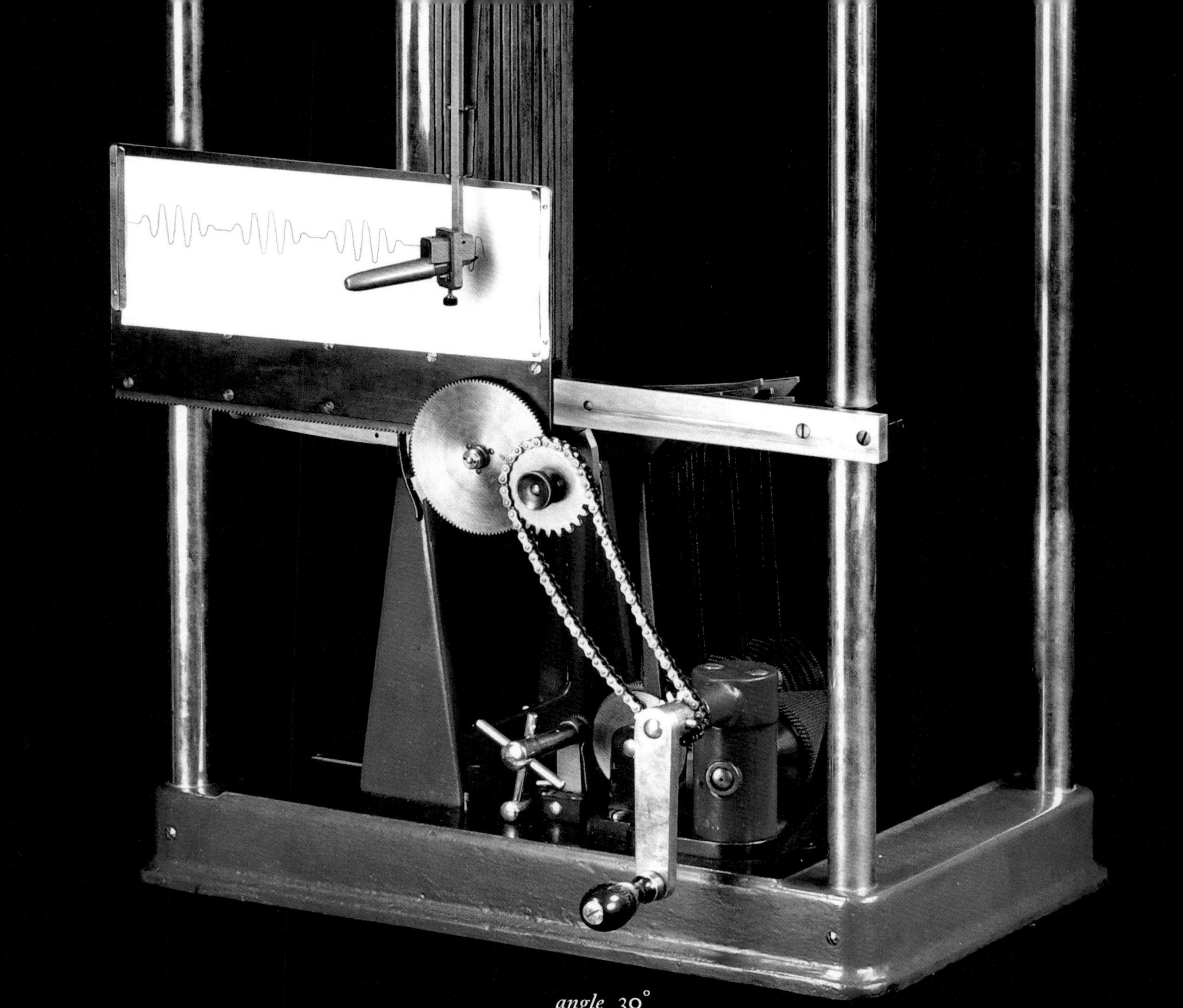

angle 30°
view 2/8

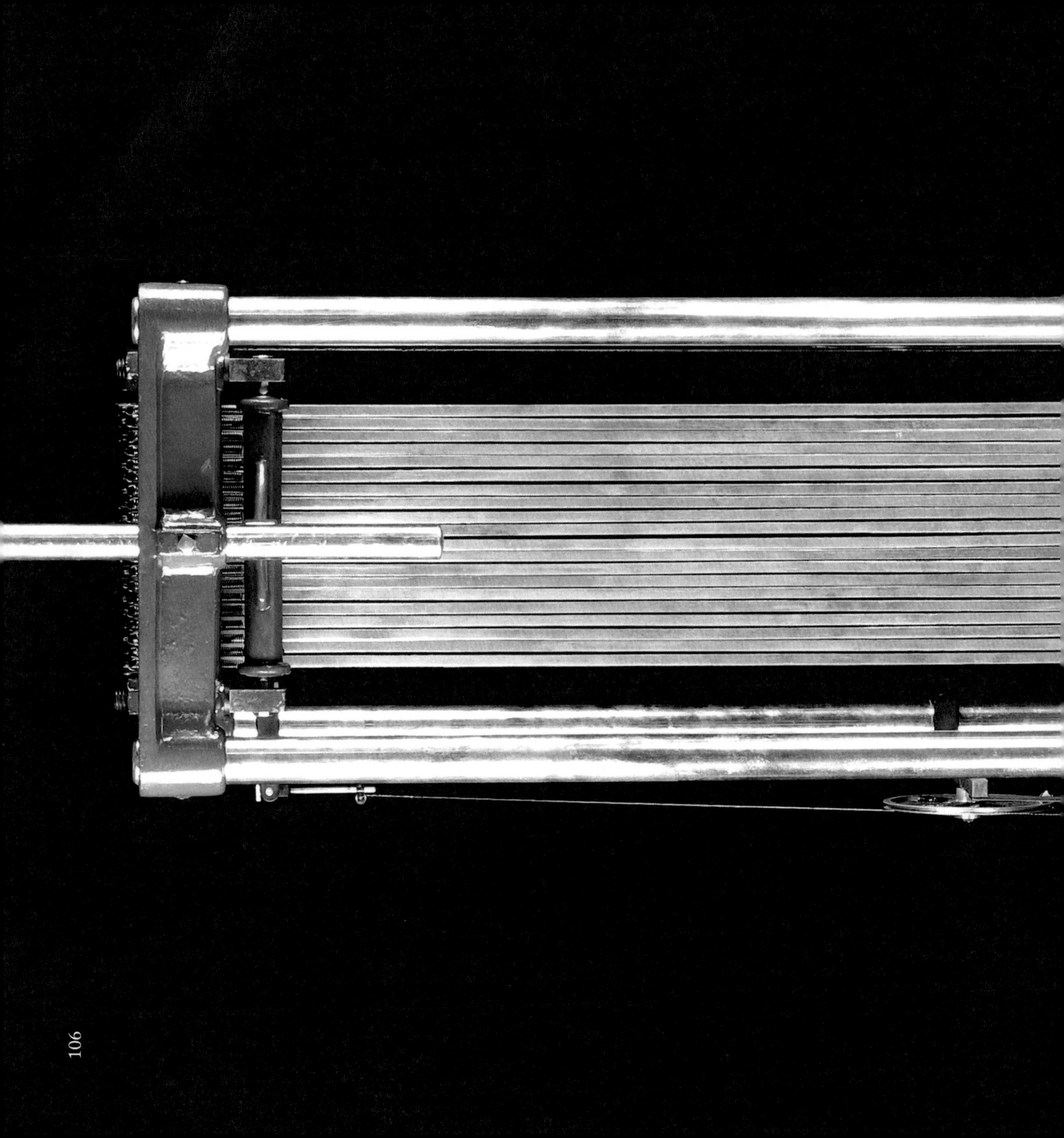

angle 90°
view 3/8

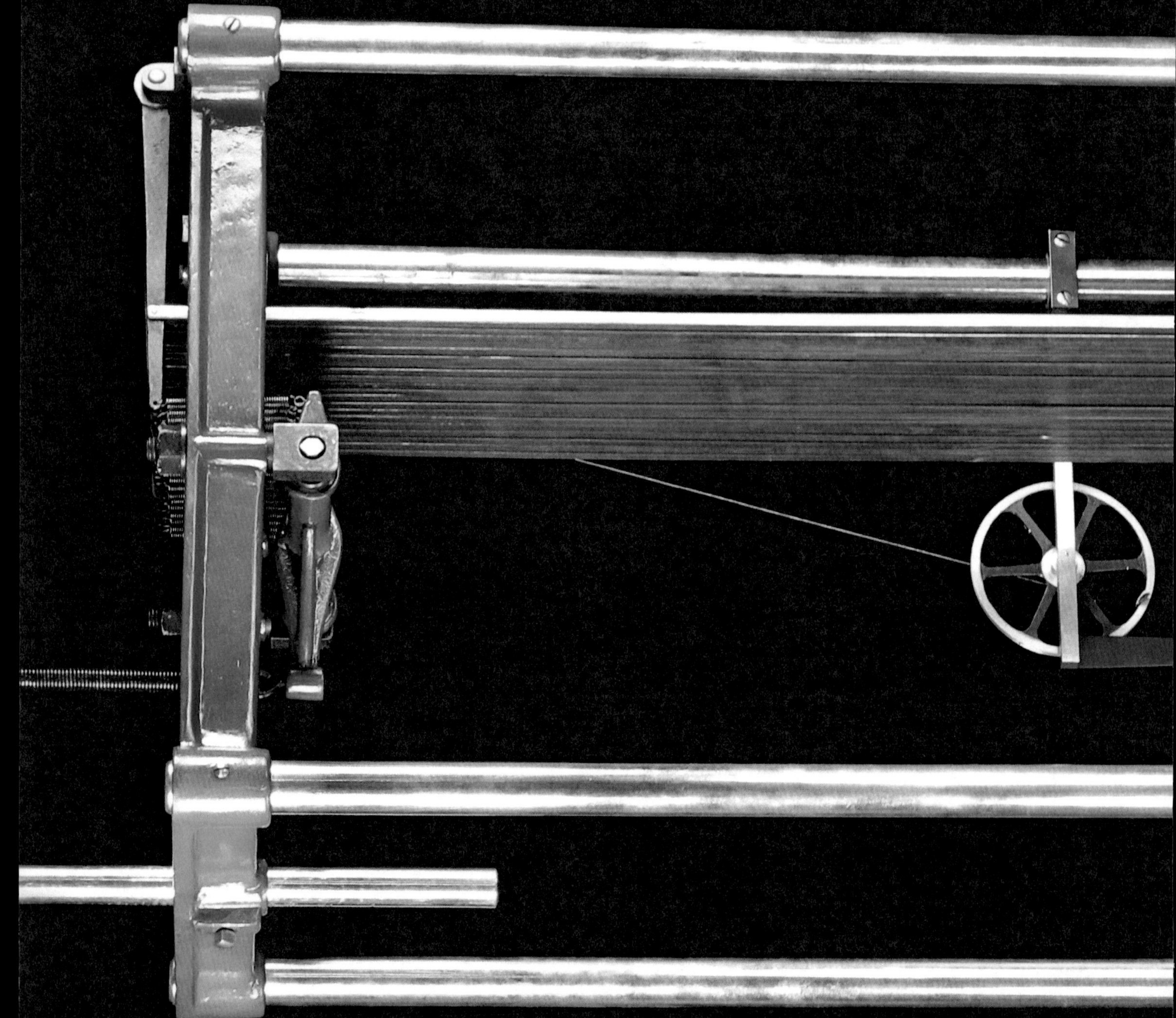

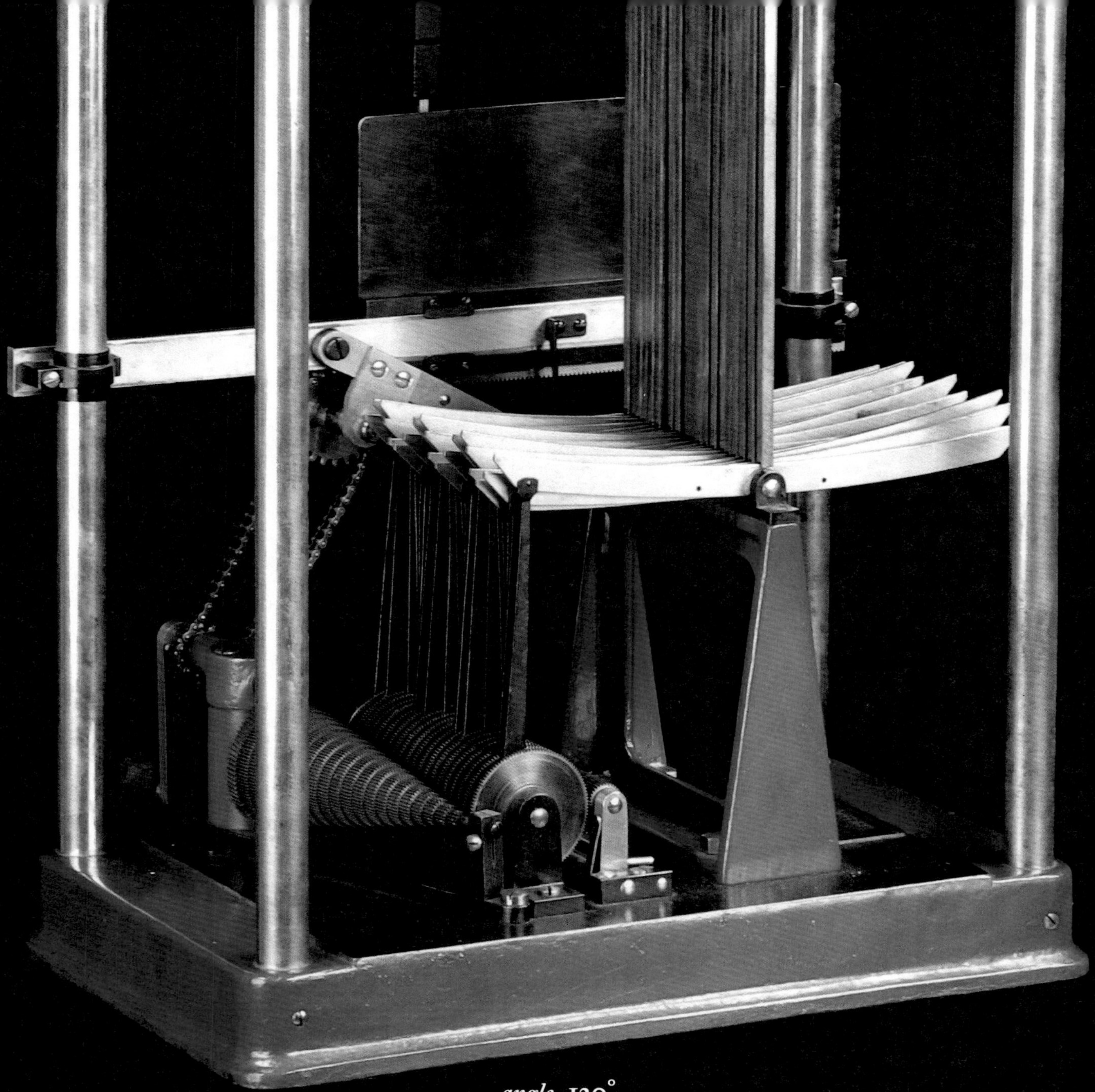

angle 120°
view 4/8

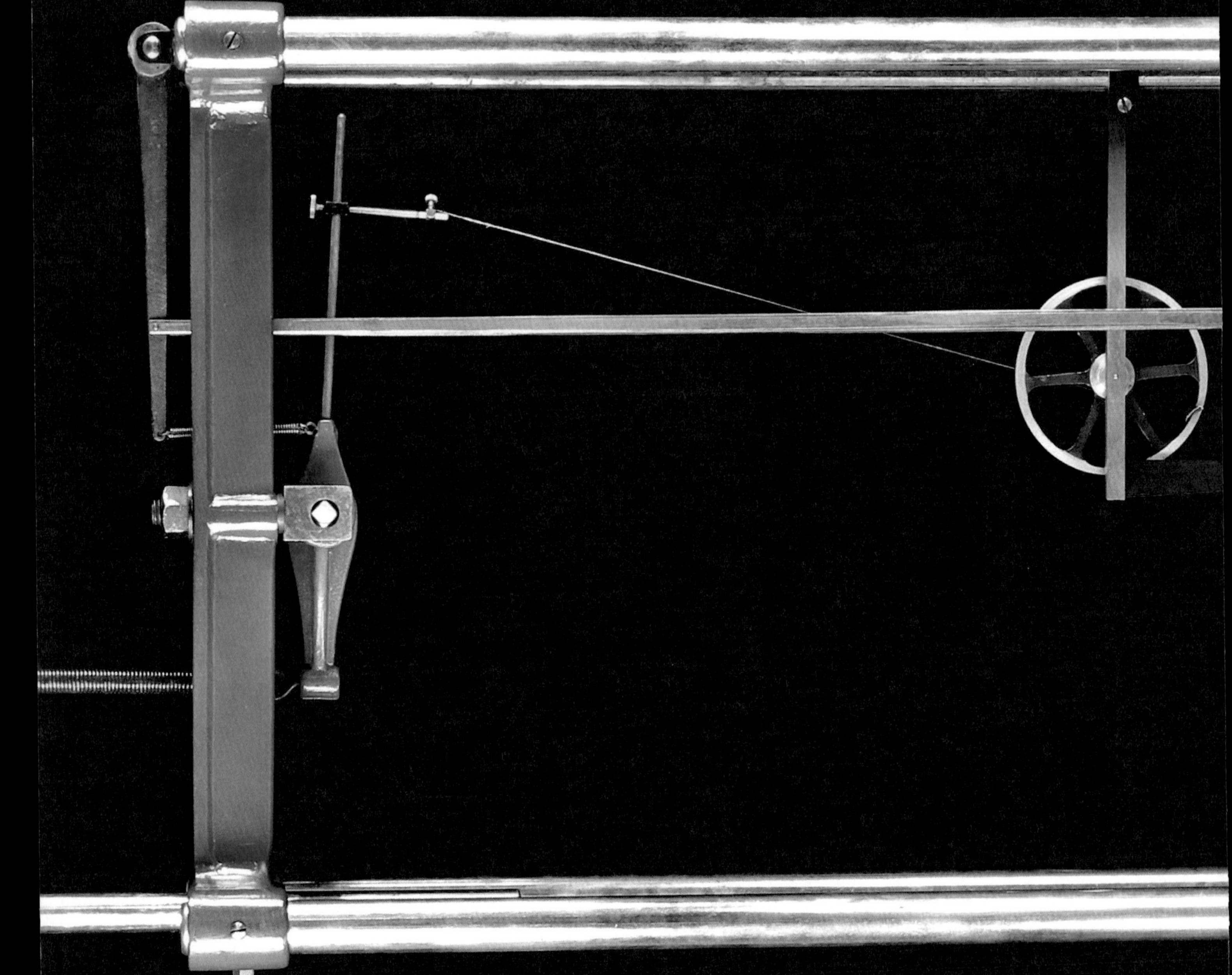

angle 180°
view 5/8

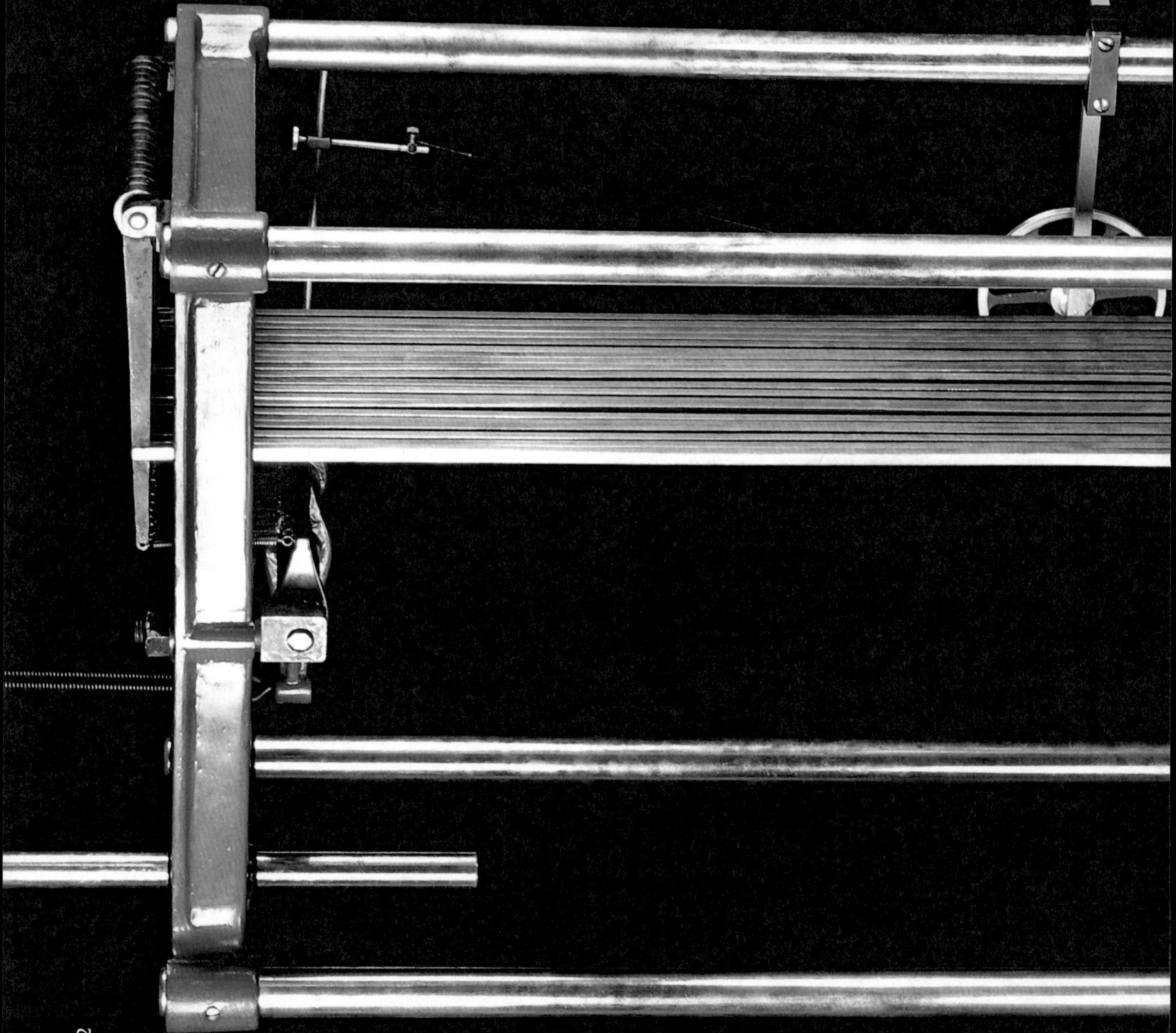

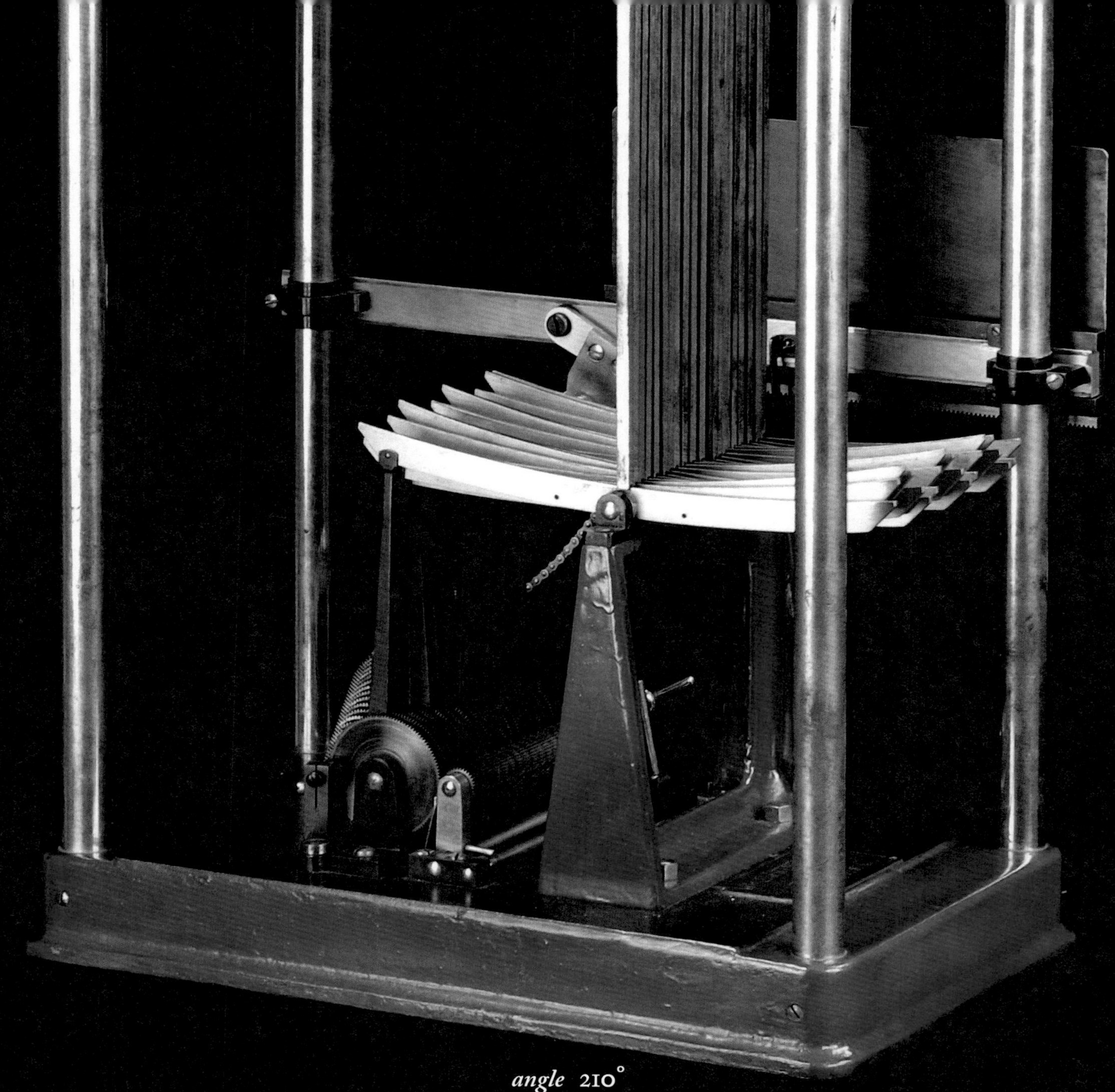

angle 210°
view 6/8

angle 270°
view 7/8

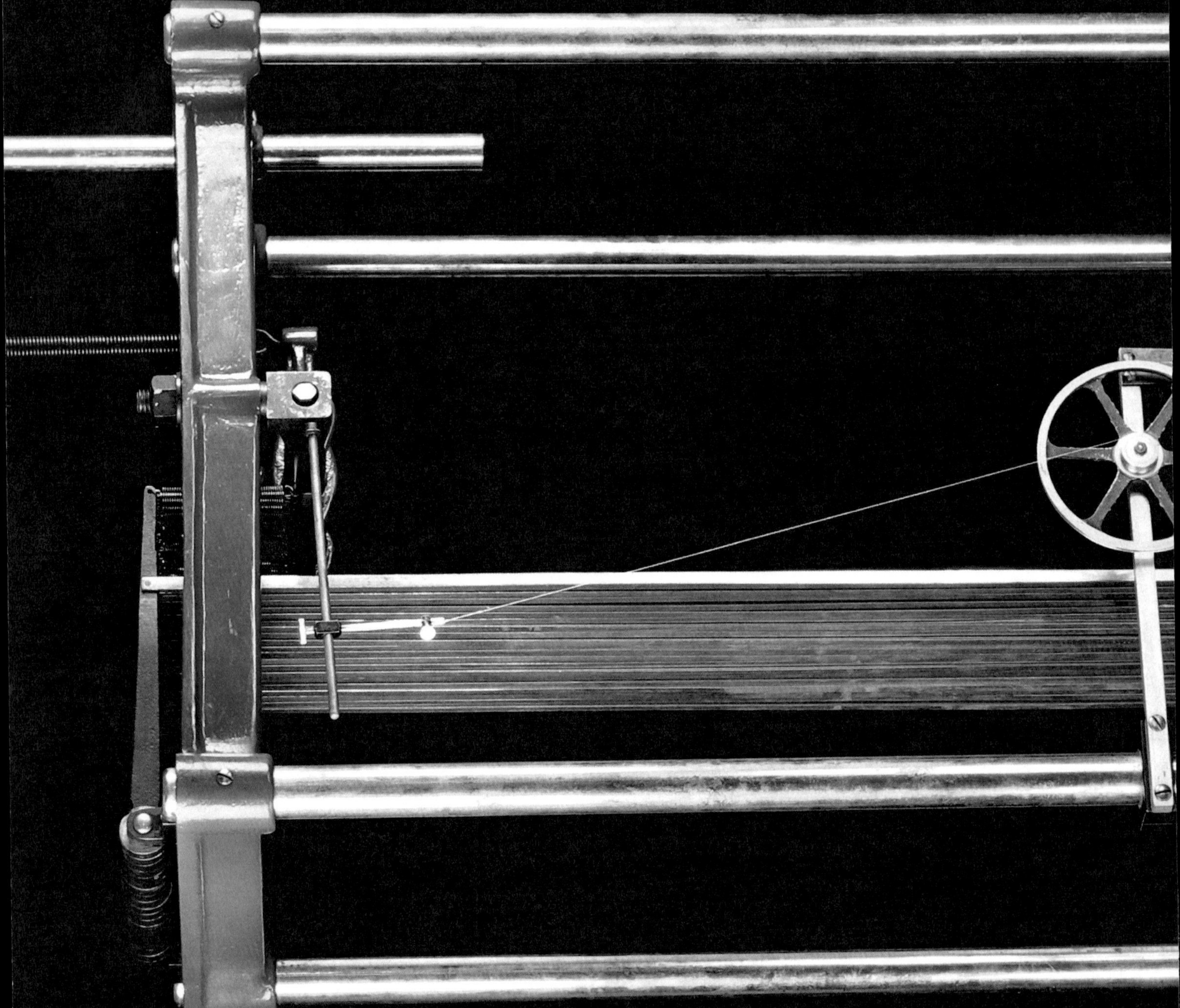

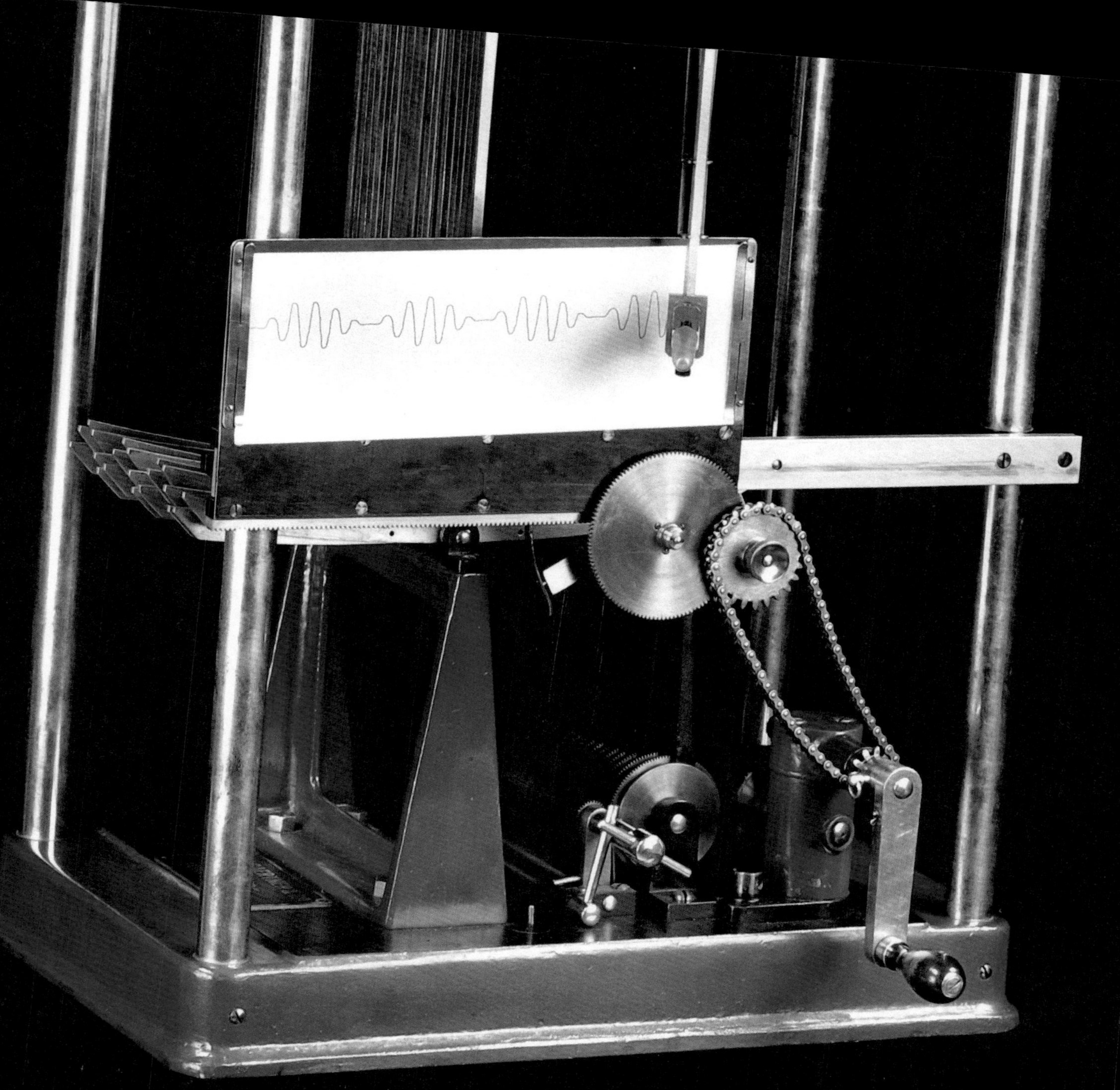

angle 300°
view 8/8

Notes on the design

Nearly all of the photographs in this book were taken using a Nikon D60 dSLR with a Tokina 100mm f/2.8 macro lens. Included in this exception are the photos on this page which were photographed using a Canon AE-1 Program with a Vivitar 20mm f/3.8 lens on Velvia 100 color slide film.

The serif text in this book is set in Adobe Caslon, the sans-serif is set in Avenir, and the title is set in Archer.

This book was laid out in Adobe InDesign.

Made in United States
North Haven, CT
25 April 2023

35856114R00073